SOLAR SYSTEM FOR KIDS

SOLAR SYSTEM FOR KIDS

A JUNIOR SCIENTIST'S GUIDE TO

Planets, Dwarf Planets, and
Everything Circling Our Sun

HILARY STATUM

ROCKRIDGE
PRESS

For general information on our other products and services or to obtain technical support, please contact our Customer Care Department within the United States at (866) 744-2665, or outside the United States at (510) 253-0500.

Rockridge Press publishes its books in a variety of electronic and print formats. Some content that appears in print may not be available in electronic books, and vice versa.

Cover Designer: Jennifer Hsu
Interior Designer: Richard Tapp
Art Producer: Tom Hood
Editor: Mary Colgan
Production Manager: Giraud Lorber
Production Editor: Sigi Nacson

ISBN: Print 978-1-64611-928-8 | eBook 978-1-64611-929-5
R1

FOR ANISTON AND HEIDI

Never stop exploring this incredible world
we've been blessed with. Much like the universe,
my love for you is inconceivable.

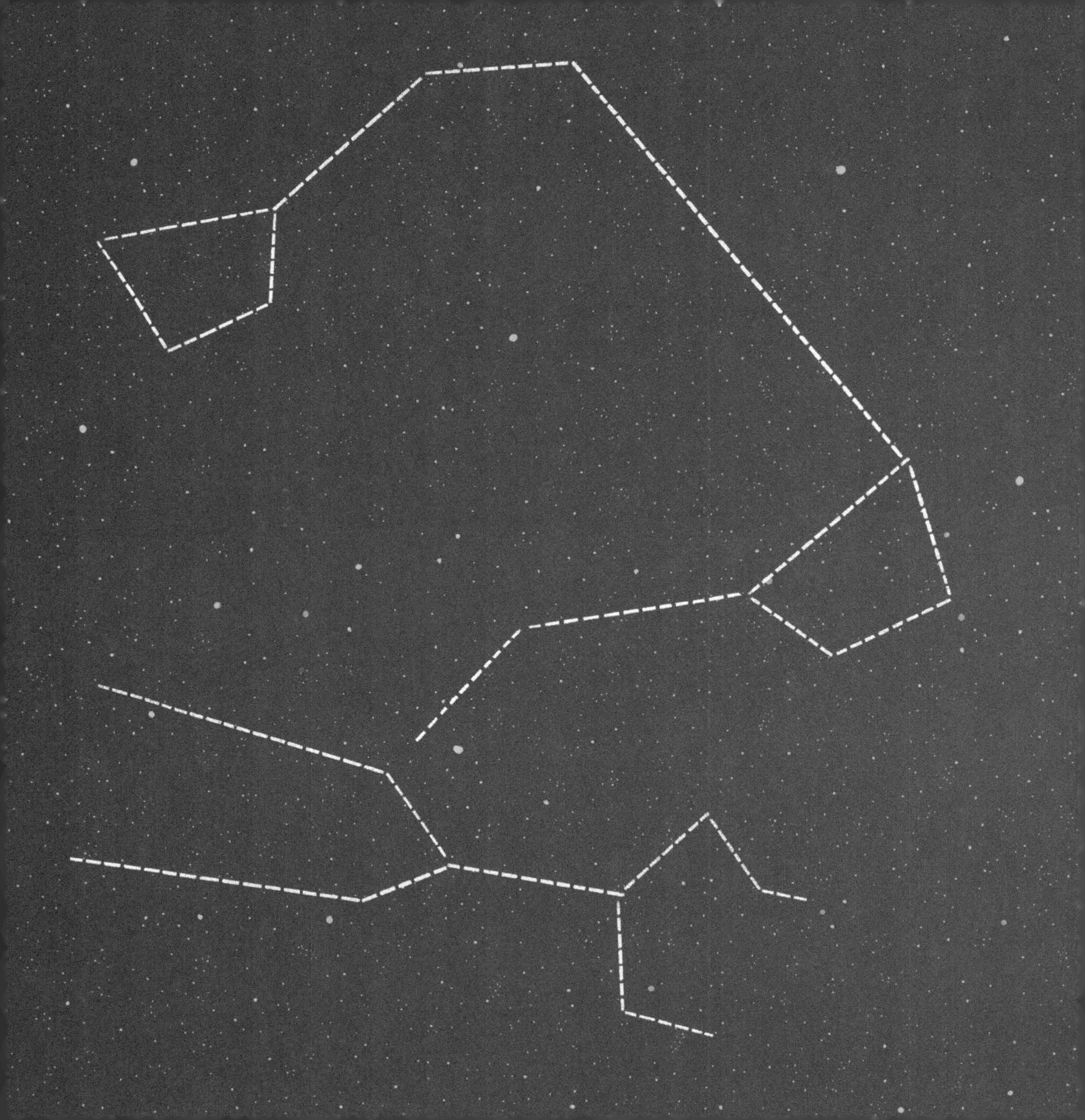

Contents

Welcome, Junior Scientist!

Have you ever looked at the night sky and wondered what is out there? Do you spend time looking for patterns in the stars? Do you wonder if aliens exist? Maybe you'd like to be an astronaut or visit another planet someday?

Well, buckle your rocket-ship seat belt and get ready to blast off. We are taking an exciting journey! Space starts about 60 miles above Earth and that is where our adventure begins. The book you hold in your hands will be your guide as you learn some of the secrets of our own solar system and beyond.

We will begin with the big picture—the **universe**. You will learn about its galaxies, stars, and planets. Once we arrive in our solar system, you will learn ways to study the stars and the Moon from your own backyard. For example, do you know how to tell if you are looking at a planet or a star? The answer is just pages away!

People have always wondered about the universe. In fact, **astronomy** became the very first science thousands of years ago. Ancient people all over the world watched the sky for changes and recorded them. Then telescopes were invented. They got bigger and better over time. Today, we have telescopes that can see objects billions of light-years away! We have sent astronauts to the Moon and launched space **probes** that have discovered new stars, **moons**, and planets. We are still learning more about space today.

Are you ready to begin your adventure? It's time to leave Earth behind!

TAKE A LOOK!

This illustration shows how the artist imagines the Big Bang (see page 12).

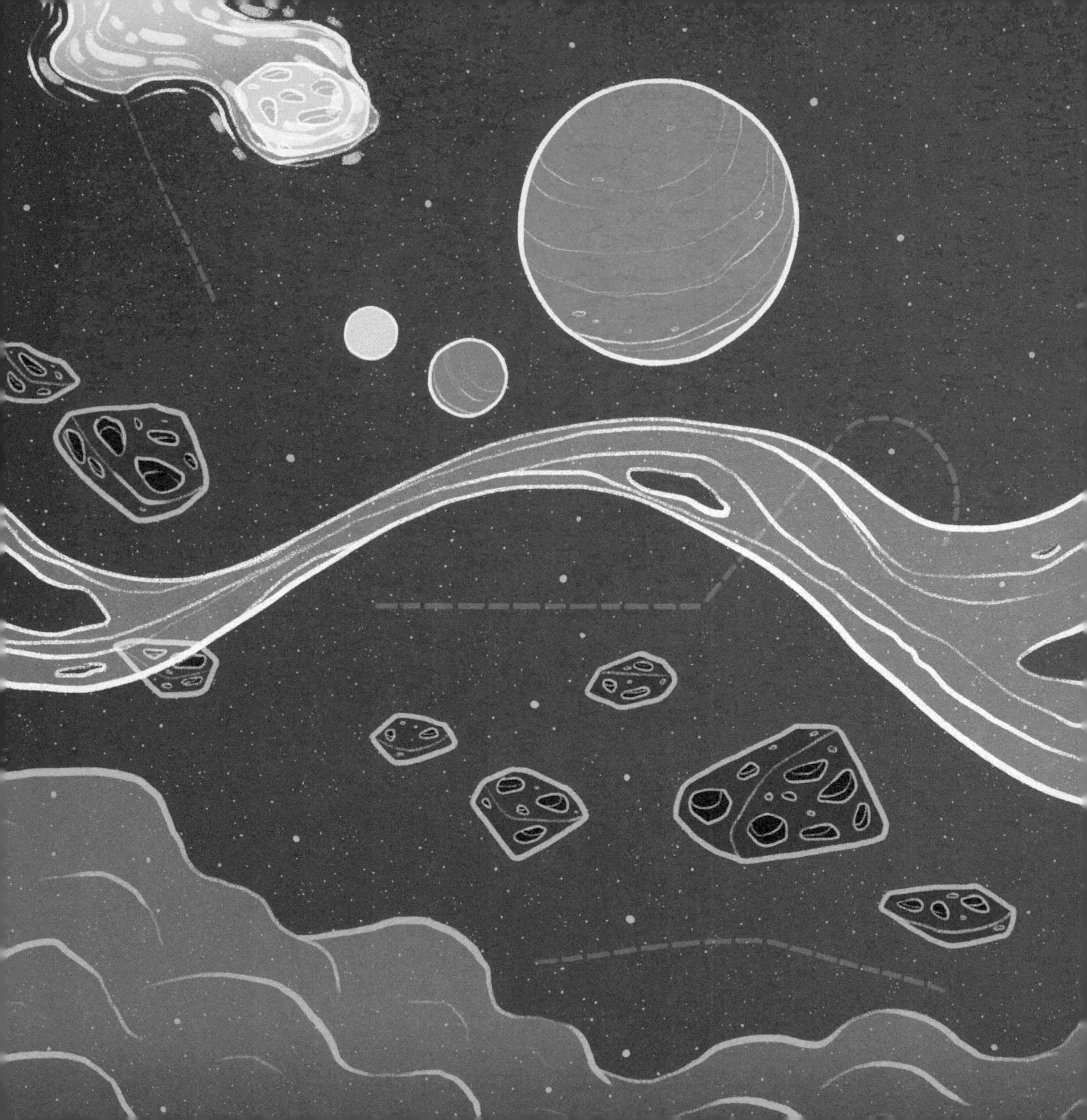

Chapter One

OUR UNIVERSE

What is the universe? That's a big question! The universe is everything: the tiniest specks of dust, the largest galaxies, and all the space in between. Trillions of planets, black holes, Earth, and our own solar system drift through what seems like endless empty space.

To give you an idea of how big the universe is, let's look at light. Light travels faster than anything else in our universe. It can travel 186,000 miles—about seven times the distance around Earth—in one second! In one hour, light can cross 670 million miles. The universe is so large that light from the farthest parts takes billions of years to reach us!

The **observable universe** is everything we can see with our eyes and telescopes. Scientists study light coming from an object to measure how fast it is moving toward or away from us. If the object is a galaxy, scientists can use this speed to figure out its distance from Earth. These measurements help scientists understand how large the universe is. Scientists think that the observable universe may reach 46 billion light-years in every direction.

There is much more to the universe than what we can see. It is always expanding, or getting bigger. This makes it difficult for scientists to figure out how large the universe really is, but that doesn't stop them from trying! Some scientists believe that the universe goes on forever!

A Big Bang

Have you ever wondered where the universe came from? It was born so long ago that no one knows for sure how it all happened. Many scientists believe that the universe began with a huge explosion about 13.8 billion years ago. Two events happened at the moment of the explosion. First, the material that would become everything in the universe came into being. Second, the universe began moving outward in all directions. This is called the Big Bang theory. A theory is an explanation for how or why things happen.

Before the Big Bang, the universe was tinier than the point of a needle.

Within one second, the universe expanded very quickly—spreading hot, energetic gas billions of miles in every direction. As all of that matter expanded and cooled, some of it began to clump together into enormous gas clouds. When these clouds came together, the first **stars** were born. It took 300 million years for the first stars to form, and our solar system didn't appear until more than four billion years later!

The force of the Big Bang was so strong that every object in the universe is still moving! Scientists think this is probably why the universe is still expanding today—and getting larger every second!

A DEEPER LOOK

Scientists say that the universe is still expanding, but how do they know? Using powerful telescopes, astronomers can see that other galaxies are moving away from our galaxy. Galaxies that are farther from us are moving away faster than galaxies that are closer to us.

Try this experiment to help you understand how this works. Blow up a balloon just a little bit. Hold it closed, and draw dots on the balloon with a marker. Then blow more air into the balloon. You will see the dots move farther away from one another. This shows how the galaxies move apart as our universe expands.

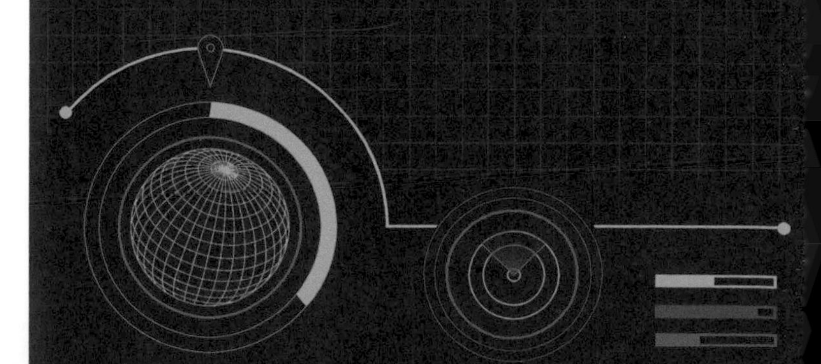

Galaxies

A **galaxy** is a group of billions of stars held together by **gravity**. Stars aren't the only things inside galaxies—they also have a lot of gas , dust, and invisible material called **dark matter**.

There are too many galaxies in the universe to count. In 1990, the Hubble Space Telescope was launched into space. The Hubble takes pictures of planets, stars, and other galaxies. At first, scientists believed there were more than 200 billion galaxies. But that changed in 2016. Photographs from the Hubble Space Telescope were used to put together a 3-D map of the universe. That map showed scientists that the observable universe has more like two trillion galaxies! That is a huge number. Two trillion is written with a "2" followed by 12 zeros (2,000,000,000,000)! The next great space telescope, the James Webb Space Telescope, will tell us more about galaxies than we have ever known.

The size, shape, age, and even the color of galaxies can be very different from one another. Scientists think that most large galaxies have a **black hole** in the middle. A black hole is an area with very strong gravity. Some galaxies are shaped like pinwheels with curved arms. Others can look like stretched balls or blobs. Galaxies have four basic shapes: spiral, barred spiral, elliptical, and irregular.

A DEEPER LOOK

Ordinary matter takes up space and has mass, or weight. It makes up all the things we can see, like trees, rocks, and people. It also makes up stars and planets. Dark matter is different. It is invisible but we know it's there because it has gravity that pulls on stars and galaxies. This mysterious stuff in space makes up about 27 percent of the universe.

Scientists know that the universe is still expanding and the rate of expansion is speeding up. They also know that there must be energy in space that is making this happen. Scientists call this dark energy.

Dark energy makes up about 68 percent of the universe. If you add these two numbers together, you will see that 95 percent of our universe is made up of dark matter and dark energy. Only 5 percent of the entire universe is made of the objects you can actually see!

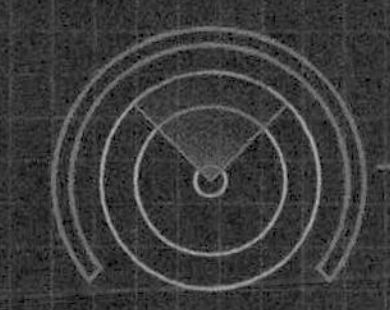

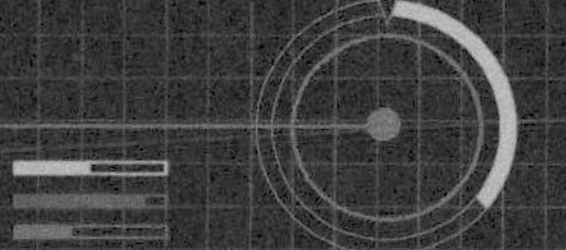

SPIRAL GALAXIES

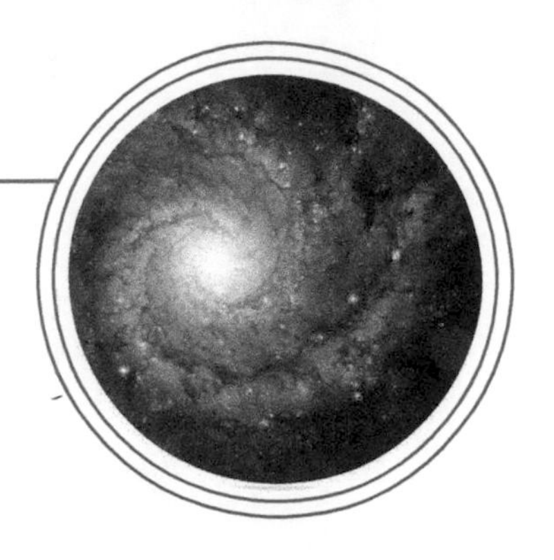

A spiral galaxy is shaped like a disk with arms that spread out from its center and circle around it like a pinwheel. Stars in the "arms" move around the spiral galaxy's core, or center. Spiral galaxies have many young, hot stars, as well as lots of gas and dust. Most spiral galaxies have a ball-shaped clump of stars in the center called a "bulge." The bulge can be seen both above and below the disk shape.

BARRED SPIRAL GALAXIES

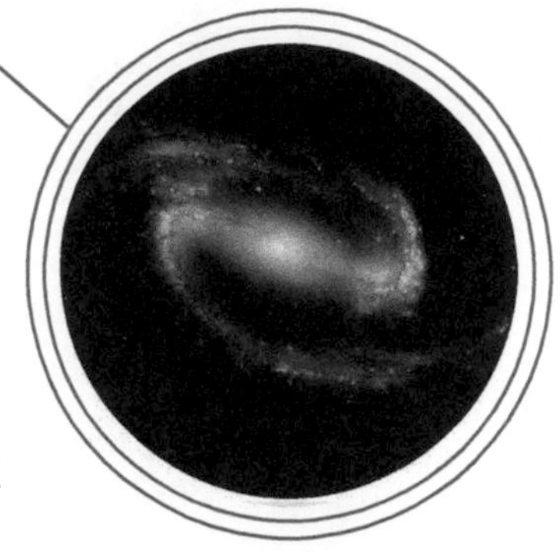

Just like its name says, a barred spiral galaxy is a spiral galaxy with a bar shape in the middle. The bar is made up of stars. About half of spiral galaxies have bars. Some bars are short, whereas others stretch across the entire galaxy. Some of these galaxies have curling arms that start from the ends of these bars. Scientists see more bars now than they did long ago, which makes them believe that bars appear as galaxies get older.

ELLIPTICAL GALAXIES

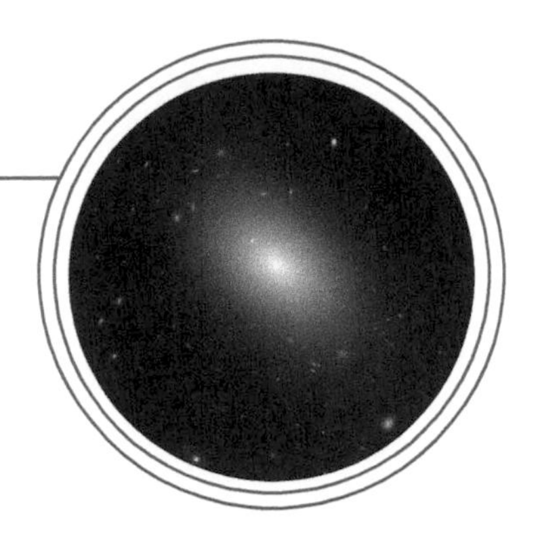

Elliptical galaxies look like stretched-out balls. Their stars seem to swarm in every direction. These galaxies come in many different sizes. Some are small, but the biggest elliptical galaxies are some of the largest galaxies we can see in the universe. Elliptical galaxies can swallow other elliptical galaxies to become even bigger! Not very many new stars are formed in these galaxies, so they contain mostly old stars. Elliptical galaxies do not contain very much gas or dust.

IRREGULAR GALAXIES

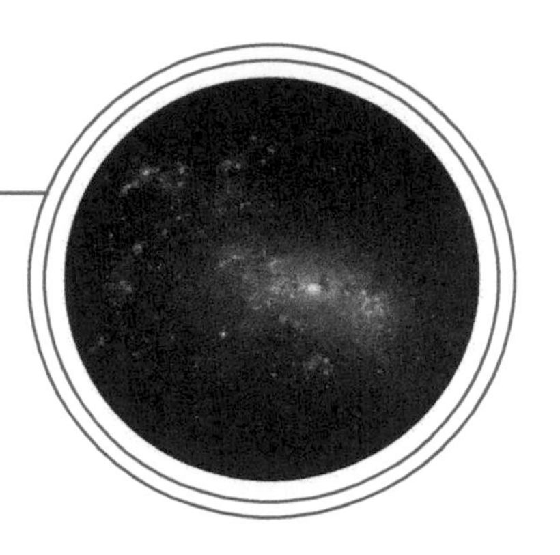

Irregular galaxies don't have a distinct shape like spiral or elliptical galaxies. Only about 20 percent of galaxies in the universe are irregular. They are some of the smallest galaxies—made up of a messy collection of stars, gas, and dust. All of that gas and dust means that irregular galaxies are often forming new stars. This means irregular galaxies are very bright.

THE MILKY WAY

Our galaxy is called the Milky Way. Why do you think it has that name? Because to many people it looks like someone has spilled a milky stripe across space!

The Milky Way is about 13.6 billion years old. It started to form shortly after the Big Bang. Our galaxy is part of a cluster of about 50 galaxies called the Local Group. The Andromeda Galaxy is the largest in the group. You can sometimes see it without a telescope!

Galaxies are many different sizes. Some have trillions of stars, whereas others have only as few as 100,000 stars. The Milky Way has around 200 billion stars. Astronomers measure distance in space using **light-years**. One light-year is equal to how far light would travel through space in one Earth year. That distance is 5.8 trillion miles! Small galaxies are a few thousand light-years across. Large galaxies can be more than one million light-years across. The Milky Way galaxy is around 100,000 light-years across.

The Milky Way has a black hole in its center. Black holes are some of the strangest things in our universe. The Milky Way's black hole is called Sagittarius A. Its gravity is very strong and is able to pull apart stars or other objects that get too close.

Our solar system is a barred spiral. It is located in one of the Milky Way's arms. On a dark night, you can see a light band of stars that stretches across the sky. That's the Milky Way! Every star you see belongs to our galaxy.

IMPORTANT PEOPLE IN SPACE HISTORY

NEIL ARMSTRONG: On July 20, 1969, this American astronaut became the first person to walk on the Moon and see Earth from space. Neil later became a professor of engineering.

MARGARET BURBIDGE: An American astronomer, Margaret helped us understand the mass and rotation of galaxies. She also designed some parts of the Hubble Space Telescope.

ANNIE JUMP CANNON: An American astronomer, Annie came up with a way to group stars based on their brightness. In 1925, she became the first woman to receive an honorary doctorate from Oxford University in England.

YURI GAGARIN: On April 12, 1961, this Russian cosmonaut (the Russian name for astronaut) became the first person to go into space. Yuri's orbit of Earth lasted less than two hours.

GALILEO GALILEI: This Italian mathematician was the first person to use a telescope to look at the night sky. Galileo made many discoveries, including Jupiter's four large moons and the craters on Earth's Moon.

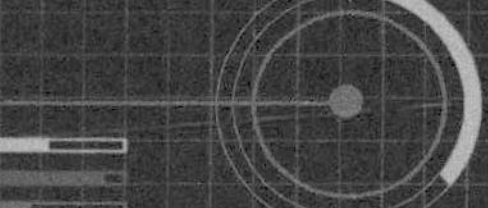

Chapter Two

OUR SOLAR SYSTEM

Now that we've learned about our home galaxy, let's travel to our home solar system! Where will we find it? Think of the Milky Way galaxy as a huge doughnut. At the center of this doughnut is the black hole. Our solar system is between the hole and the outer edge of the doughnut. It is about 25,000 light-years from the black hole and 25,000 light-years from the edge. Our Sun is located on a small arm of the Milky Way.

Even though our solar system is very large, it is tiny compared with the size of the Milky Way galaxy. Remember, the Milky Way is about 100,000 light-years across. Our solar system is only about one light-year across. Our Sun looks like a tiny little speck among billions of other stars in the galaxy.

So, what is a **solar system**? A star with **planets** that **orbit**, or move around, it is called a **planetary system**. The term *solar system* is used only to describe our planetary system. There are many planetary systems in our galaxy. Scientists have found more than 2,800 other stars in the Milky Way galaxy with planets orbiting them.

The Sun is a star at the center of our solar system. In fact, the word *solar* means "sun." Eight planets orbit the Sun: Mercury, Venus, Earth, Mars, Jupiter, Saturn, Uranus, and Neptune. Other objects also orbit the Sun. They include asteroids, dwarf planets, and comets. The Sun is so massive that its gravity pulls on each planet and everything else nearby. This gravity keeps all of the smaller objects moving around it. Earth takes one year to orbit the Sun. Planets that are closer to the Sun take less time to orbit, whereas planets that are farther away take longer.

How did our solar system form? Many scientists think our solar system formed about five billion years ago and began as a giant cloud of dust and gas. The Sun formed from the dust and gas first. Then, dust particles from the cloud swirled around the new Sun. These dust particles started to bump into one another and stick together. Eventually, clumps of dust and **matter** became planets, comets, asteroids, dwarf planets, and meteoroids. We will learn all about each of these things later in the book.

What Is a Star?

Stars are glowing balls of burning gases. The Sun is the only star in our solar system. There are about 200 billion stars in the Milky Way galaxy, but you can see only about 5,000 of them from Earth. Many stars look white in the night sky, but they can be many different colors. The hottest stars are blue, and the coolest ones are red. In between these extremes are yellow and orange stars, which look closer to white to our eyes.

Did you know that a shooting star is not really a star? It is actually a streak of light called a **meteor** that is caused by a small chunk of rock burning up in the Earth's atmosphere!

The Life Cycle of a Star

Stars are created in a large cloud of dust particles and gas. This cloud is called a **nebula**. Dust and gas clump together into a cloudy ball. This ball is a newborn star called a protostar. A star is born!

The middle of the protostar gets hotter and hotter over time. The center of the star is made of a gas called hydrogen. As the center of the protostar gets hotter and denser, a process called fusion causes the hydrogen to turn into helium gas. This change makes a lot of energy. The energy gives off light and the star begins to shine. The star is now an adult.

An adult star, including the Sun, can burn for billions of years. But eventually, its hydrogen will begin to run out. When this happens, the star starts to use helium as a fuel. This causes it to cool, expand, and become a brighter red giant star. More massive stars have to cool and expand more than Sun-like stars to shine even brighter. That is why they are called red supergiants!

A star's mass seals its fate. If a star is more than about eight times as massive as the Sun, it will become a red supergiant and explode at the end of its life. The explosion is called a supernova. It happens when the star runs out of fuel. After the supernova, the star collapses in on itself. It becomes either a neutron star or a black hole.

Less massive stars will not explode after becoming a red giant. Instead, they toss off their outer layers to form a nebula while the core shrinks into a white dwarf. A white dwarf can shine for billions of years, but it's not powered by fusion anymore, so it slowly fades. Black dwarf stars are those that are at the very end of their life cycles and don't make heat or light.

A brown dwarf is heavier than a giant planet but lighter than a small star. Brown dwarfs are sometimes called failed stars because they are not large enough to turn hydrogen into helium. They give off very little light.

During their lives, stars change temperature and size as they burn up their hydrogen supply. The biggest stars have the shortest lives. Still, even giant stars live for a very long time. Stars live anywhere from a few million years to billions of years!

TAKE A LOOK!

Artists sometimes use their imaginations and play with shapes and colors when drawing something real. You can find photos of different kinds of stars at nasa.gov/kidsclub. How would you draw a star's life cycle?

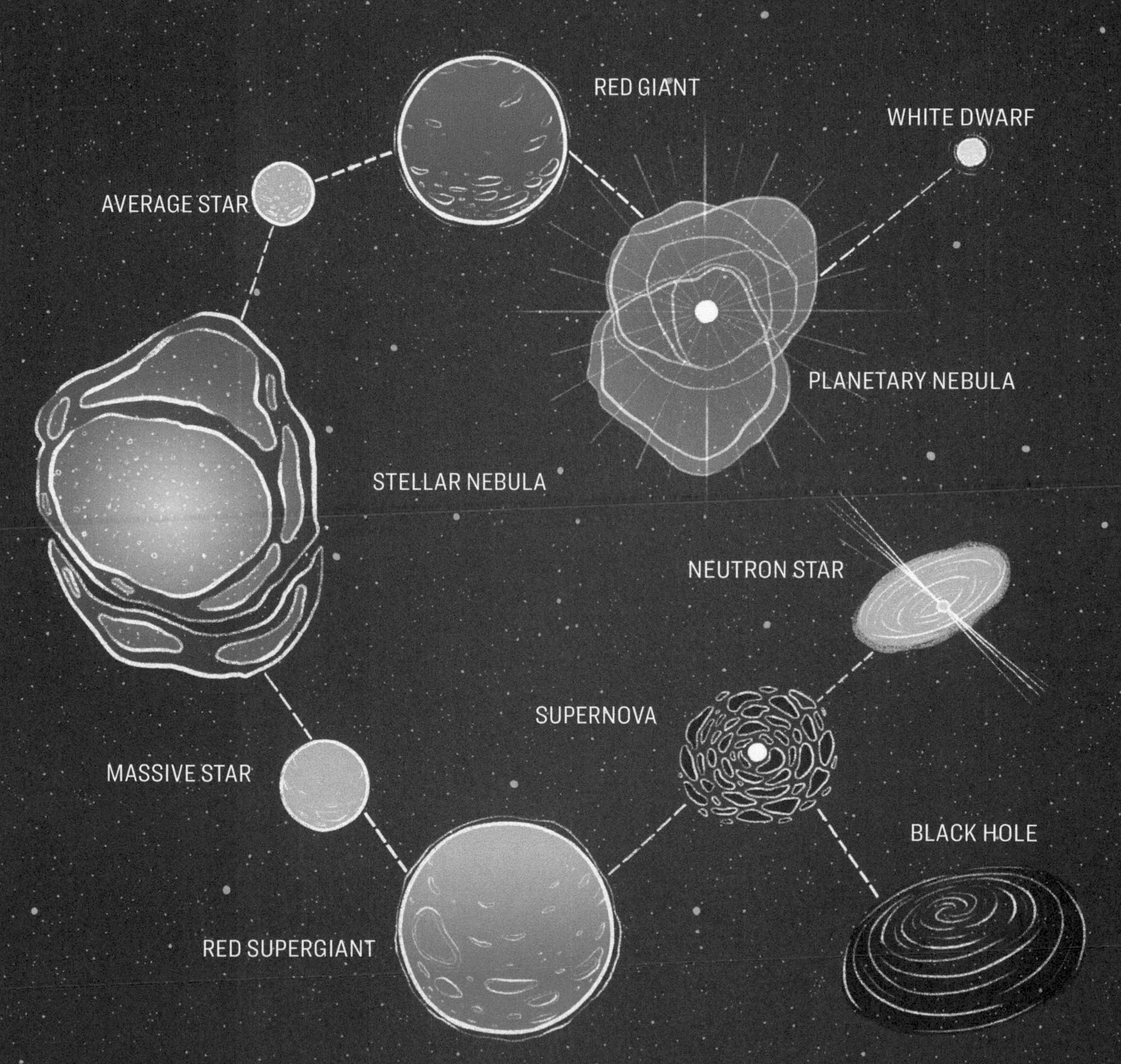
LIFE CYCLE OF A STAR
RED GIANT
WHITE DWARF
AVERAGE STAR
PLANETARY NEBULA
STELLAR NEBULA
NEUTRON STAR
SUPERNOVA
MASSIVE STAR
BLACK HOLE
RED SUPERGIANT

THE SUN

MASS:
1.989 X 10^{30} (333,000 EARTHS!)

DIAMETER:
864,337 MILES

DISTANCE FROM EARTH:
93 MILLION MILES

AVERAGE TEMPERATURE:
27 MILLION DEGREES FAHRENHEIT

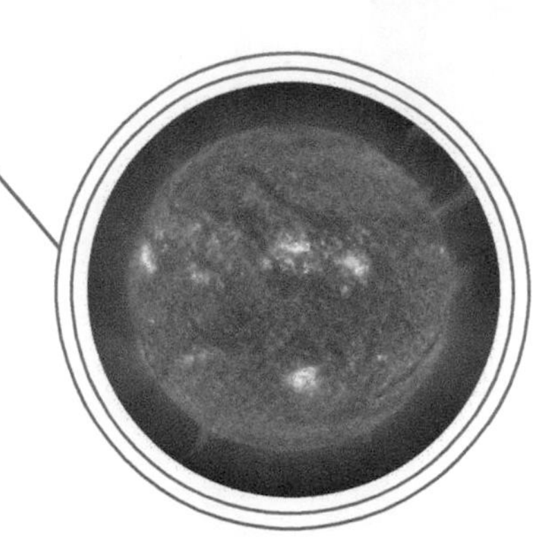

Our Sun is the closest star to Earth. It is a yellow dwarf star, which means it is an average temperature. The Sun is the largest object in our solar system. About 1.3 million planet Earths could fit inside it!

The Sun's gassy surface is always changing. Sometimes, the Sun throws giant flaming loops of gas into space. These are called **prominences**. Other times, the Sun shoots large explosions of energy into space. These explosions are called solar flares. The Sun's hot gasses constantly bubble to its surface, then sink back down again as they cool. The cooler, darker areas on the Sun's surface are called sunspots. Some sunspots are bigger than our planet!

Every planet in our solar system orbits the Sun, but did you know the Sun is moving, too? Our entire solar system travels around the center of the Milky Way galaxy. It takes our solar system about 225 million years to orbit the Milky Way.

The Sun is very important for all life on Earth. Plants turn sunlight into the energy they need to grow. Humans and animals depend on plants for their food. Without the Sun, the lack of food wouldn't be the biggest problem. Earth would be freezing cold—much too cold for anything (or anyone) to live.

After learning about the life cycle of stars, you may be wondering if the Sun can last forever. It can't. But thankfully the Sun will continue to be around for a very long time. Scientists think that in about five billion years the Sun will run out of hydrogen. Like other stars, the Sun will grow and become a red giant, many times brighter than it currently is. It will become large enough to swallow nearby planets, including Earth. After another 100 million years, the Sun will shrink and turn into a white dwarf star.

Our Sun is amazing, and you would probably love to take a good look at it. But remember to never look directly at the Sun. The Sun is so bright it can badly damage your eyes.

A DEEPER LOOK

Many years ago, some people believed the Sun orbited Earth. In 1543, that idea changed. Nicolaus Copernicus proposed a different model with the Earth moving around the Sun. Later, in 1610, Galileo Galilei spent many years studying the Moon's phases, Jupiter and its moons, and other planets. When Galileo pointed his new telescope toward Venus, he was amazed to see the planet go through phases just like the Moon does. He correctly suggested that this could happen only if Venus had an orbit between Earth and the Sun.

Today, NASA (National Aeronautics and Space Administration) has a fleet of spacecraft to observe the Sun and the objects that orbit it.

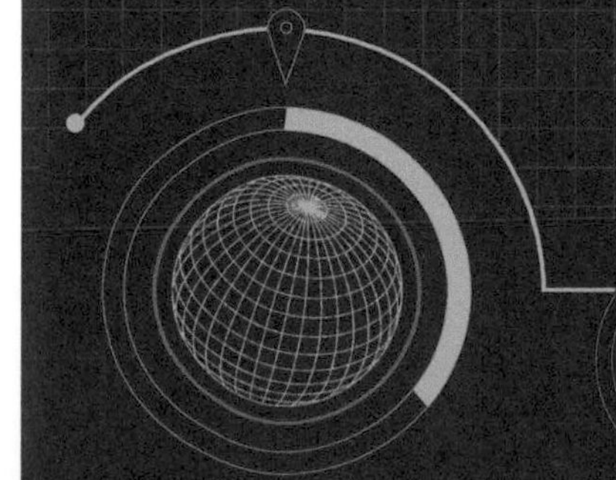

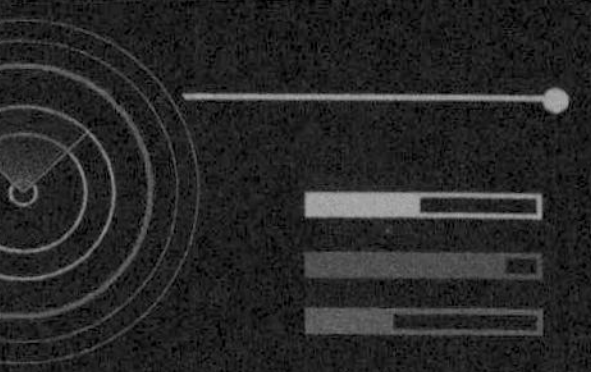

SOLAR ECLIPSES

Have you ever seen a **solar eclipse**? A solar eclipse happens when the Moon passes between Earth and the Sun, blocking some or all of the Sun's light. If only a part of the Sun is blocked, it is called a partial solar eclipse. If the Moon lines up perfectly between Earth and the Sun, it is called a total solar eclipse. When this happens, the Moon blocks the Sun's light almost completely!

During a solar eclipse, the Moon casts a shadow on Earth and the Sun's light becomes dim. Daytime looks more like nighttime to people on the ground. A total solar eclipse happens only every one or two years for a few minutes. Even then, people in only a small area on Earth are able to see it. It may be tempting, but never look directly at a solar eclipse as it could hurt your eyes. You can buy special sunglasses to view this rare event.

TAKE A LOOK!

This illustration shows the phases of a solar eclipse. At the center, you can see the Moon passing over the Sun in a total solar eclipse.

What Is a Planet?

A planet is a large, round object that orbits a star. In order to be called a planet, the round object must be big enough to clear away any other objects in its path around the star. A **dwarf planet** is not large enough to clear its own pathway. It orbits a star following the same route as other small objects, like asteroids. We will explore dwarf planets later in the book.

Our solar system has eight planets, but there are many more planets in the Milky Way galaxy. In fact, scientists think our galaxy has between 800 billion and 3.2 trillion planets—about one for every star, on average!

TAKE A LOOK!

This illustration shows orbital paths around the Sun. Draw your own with all the planets in place!

Some planets are solid and rocky like our planet. This kind of planet is called a **terrestrial planet**. Mercury, Venus, Earth, and Mars are the terrestrial planets in our solar system. Other planets, like Jupiter and Saturn, are made of gas. They are called gas giants. Uranus and Neptune are ice giants—gas giants that have large, icy centers.

Like most things in the universe, planets are on the move! You've learned that planets in our solar system are orbiting around the Sun. While they orbit, they are also rotating, or spinning, on their axes. A planet's day is the amount of time it takes to rotate, or turn, once on its **axis**. A planet's year is the time it takes to orbit the Sun once. Earth's day is 24 hours, and its year is 365 days.

In the next two chapters we will learn more about the eight planets in our solar system.

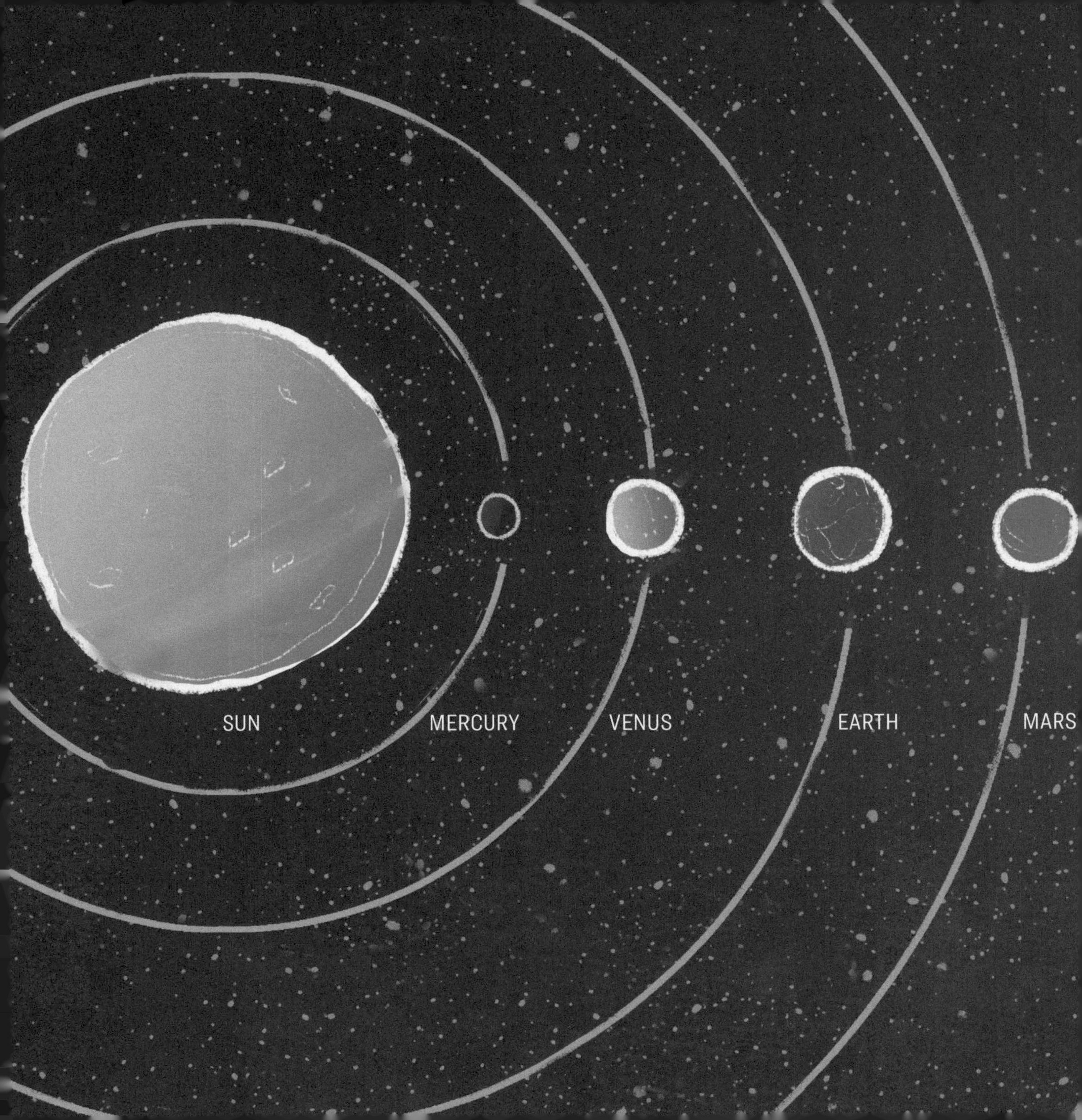

SUN
MERCURY
VENUS
EARTH
MARS

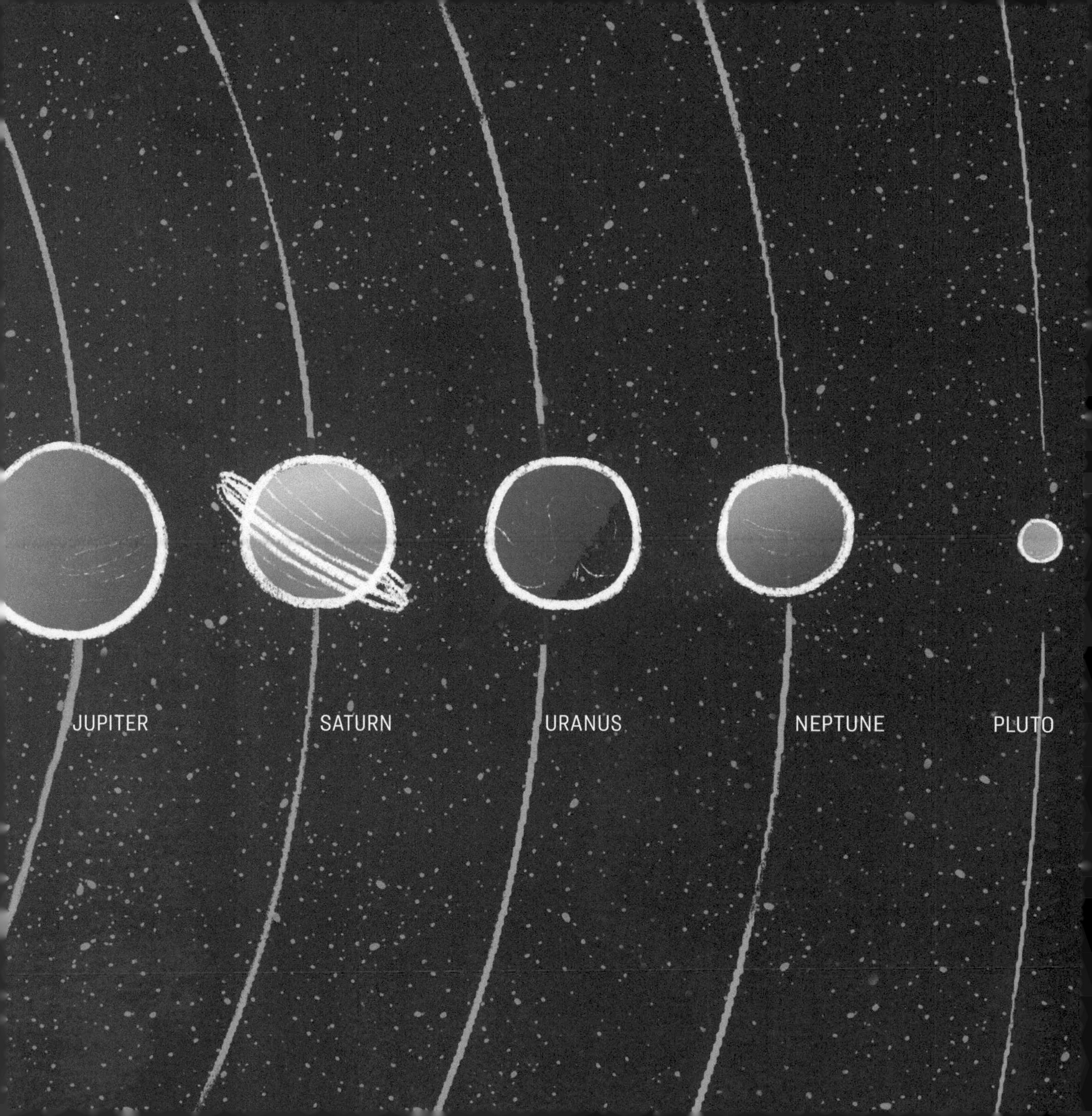

JUPITER

SATURN

URANUS

NEPTUNE

PLUTO

WHO NAMED THEM?

PLANETS

All of the planets in our solar system, except for Earth, were named after Greek or Roman gods and goddesses from mythology.

MERCURY: In Roman mythology, Mercury is the god of buying and selling, stealing, and travel. Because it's closest to the Sun, Mercury moves fast in its orbit. This quality is probably why the planet was named after this god.

VENUS: Venus is named after the Roman goddess of beauty and love. It is brighter than any other planet in the sky because it is closest to Earth and similar in size.

EARTH: Earth's name comes from the German and Old English words *erda* and *eorthe*, meaning "ground."

MARS: Mars, named after the Roman god of war, most likely got its name because of its red color.

JUPITER: Jupiter is the largest planet in our solar system. In Roman mythology, Jupiter is the king of the gods.

SATURN: Saturn is named after the Roman god of agriculture, or farming.

URANUS: Uranus is named after the ancient Greek god of the heavens.

NEPTUNE: Neptune is an amazing blue color, so it was named after the Roman god of the sea.

DWARF PLANETS

The International Astronomical Union has named five of the largest dwarf planets in our solar system; however, there may be several hundred to several thousand dwarf planets in orbit.

CERES: Ceres is named after the Roman goddess of corn and harvests.

PLUTO: This dwarf planet's name was suggested by an 11-year-old girl! Pluto is the Roman god of the underworld. Pluto is so far away from the Sun that it is always dark.

HAUMEA: In Hawaiian mythology, Haumea is the name of the goddess of childbirth and fertility. Before its official name, some scientists called this dwarf planet "Santa" since it was discovered just after Christmas in 2004!

MAKEMAKE: In Rapa Nui mythology, Makemake is the chief god, the god of fertility, and the creator of humans.

ERIS: In Greek mythology, Eris is the goddess of discord, or disagreements.

A DEEPER LOOK

Are you curious about the moons in our solar system? Giant Jupiter's moons are some of the most famous—and the largest. In 1610, Italian astronomer Galileo saw three bright objects near Jupiter. At first, he thought they were stars, but then he realized they were moons orbiting around the planet. He later found a fourth. Aside from the Sun and the other seven planets, these moons are the largest objects in our solar system. They are named Io, Europa, Ganymede, and Callisto. You can remember their order from inner to outer by thinking of the sentence "I Eat Green Carrots." Astronomers later named the group the Galilean Moons in his honor.

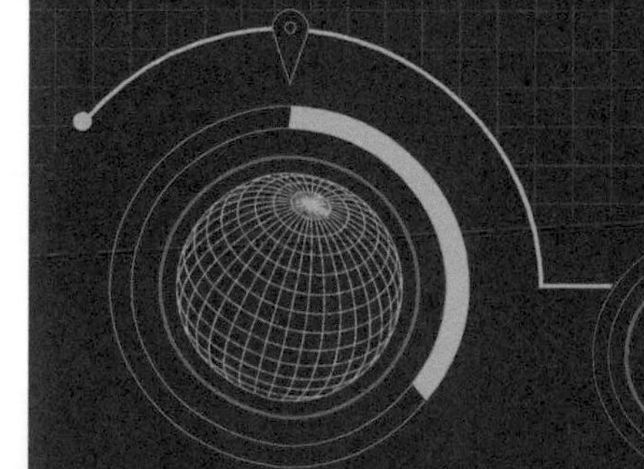

What Is a Moon?

You know that Earth has a moon, but did you know that many other planets in our solar system have moons, too? Earth's moon is only one of hundreds in the solar system. Some planets have more than one moon. Jupiter has 79! Moons are sometimes called natural satellites. A **satellite** is something that orbits a larger object. Moons are satellites of their planets, just as all of the planets are satellites of the Sun. An interesting fact is that human-made satellites are sometimes called artificial moons.

Moons are made up of chunks of rock, and many moons probably formed from the same cloud of dust and gas that the planets did. Moons can be many different shapes and sizes. They are usually solid, and many don't have their own **atmospheres**. Most smaller moons are actually asteroids that were captured by the gravity of a large planet when they passed too close. Many scientists think Earth's moon was formed when our planet was hit by a Mars-sized

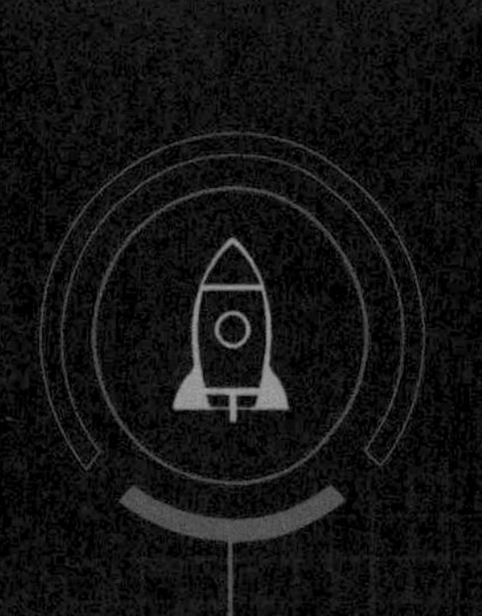

EXPLORE MORE!

You can make a shadow clock that uses Earth's rotation to tell time! Push a wooden pole in the ground on a sunny day. Each hour, mark where the shadow falls with a craft stick that has the time written on it (11:00, 12:00, etc.). When it's finished, you can use your shadow clock to tell the time!

object. The crash sent giant pieces of Earth into space. Gravity pulled the pieces together to form the Moon.

Moons travel with their planet through space. The planet's gravity holds its moons in orbit. Some asteroids even have moons. At first, scientists didn't think asteroids could have enough gravity to hold moons in their orbits. Then, in 1993, the *Galileo* spacecraft discovered a tiny moon orbiting an asteroid named Ida. The moon was later named Dactyl. We now know that a few asteroids have small moons in their orbit.

Ask your friends or family members if they know how many moons have been discovered so far in our solar system. (The answer is more than 200.) Most people are surprised by the answer!

EXPLORE MORE!

Make your own moon! Trace two circles onto a piece of aluminum foil, then carefully cut them out. Crinkle the foil to give it lots of texture, then gently smooth it out. Cut a cardboard circle the same size and glue the foil circles on the front and back. Punch a hole in the top, thread a string through the hole, and hang it from your bedroom ceiling!

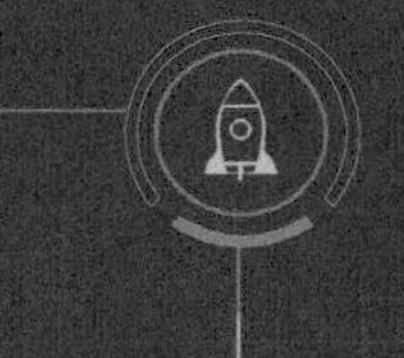

MORE IMPORTANT PEOPLE IN SPACE HISTORY

MAE JEMISON: In 1992, Mae became the first black woman in the world to go into space when she flew aboard the Space Shuttle *Endeavor*. She worked at NASA as an astronaut for six years.

KATHERINE JOHNSON: Katherine provided NASA with the necessary mathematical calculations to launch many of its most famous missions, including John Glenn's 1962 orbit of the Earth.

SIR ISAAC NEWTON: A British hero of science whose groundbreaking theory of gravity helped explain the movements of the planets around the Sun.

ELLEN OCHOA: In 1993, Ellen became the first Hispanic woman to enter space when she traveled aboard the shuttle *Discovery* for an eight-day mission.

VALENTINA TERESHKOVA: In 1963, this Russian cosmonaut became the first woman to be launched into space. Valentina spent three days orbiting Earth on her solo mission aboard the *Vostok 6*.

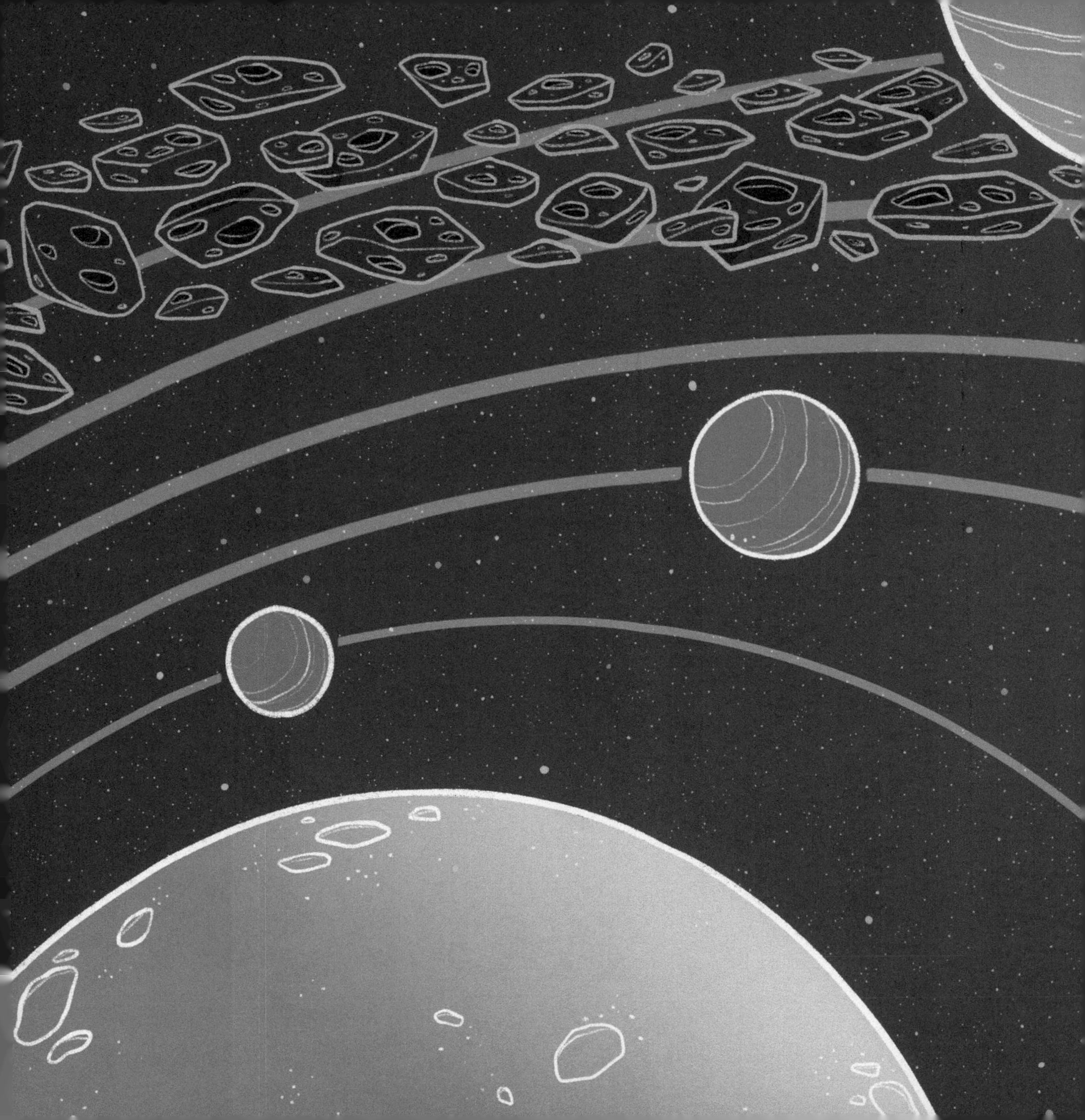

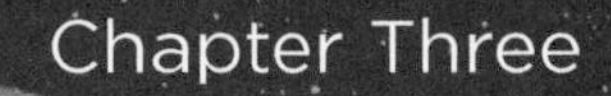

Chapter Three

TERRESTRIAL PLANETS AND THE ASTEROID BELT

Are you ready to learn about the planets in our solar system? Let's go! We will begin with the terrestrial planets. Terrestrial planets have a hard surface and a liquid rock core. They are mostly made up of rocks or metals. Terrestrial planets can have volcanoes, mountains, craters, and canyons on their surfaces. They can have thick or thin atmospheres. Terrestrial planets have no rings and few moons. These features may sound familiar. Why? Because you live on a terrestrial planet! The four planets that are closest to our Sun are the terrestrial planets in our solar system: Mercury, Venus, Earth, and Mars.

MERCURY

DIAMETER:
3,032 MILES

DISTANCE FROM SUN:
35.98 MILLION MILES

DISTANCE FROM EARTH:
48 MILLION TO 138 MILLION MILES (DEPENDING ON PLACE IN ORBIT)

KNOWN MOONS:
NONE

LENGTH OF 1 DAY:
59 EARTH DAYS

LENGTH OF 1 YEAR:
88 EARTH DAYS

AVERAGE TEMPERATURE:
332 DEGREES FAHRENHEIT (RANGES FROM 800 TO -290 DEGREES FAHRENHEIT)

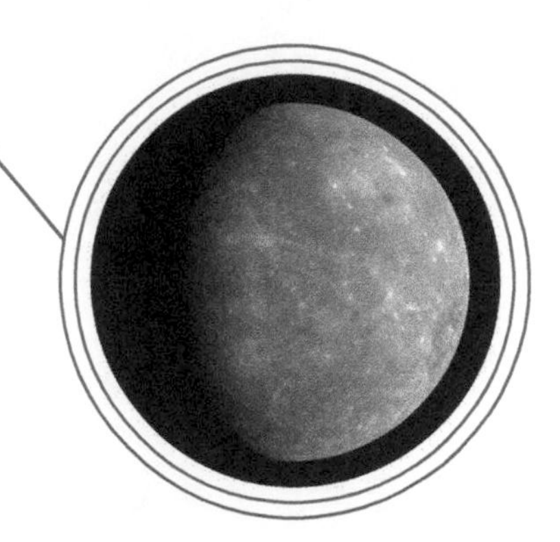

Mercury is the closest planet to the Sun and is also often Earth's closest neighbor. The days and years on Mercury are very different from Earth's. In fact, a year on Mercury takes only 88 Earth days. It takes 59 Earth days for Mercury to spin around once on its axis. Mercury's short years and long days mean that the Sun rises only once every 180 Earth days there!

Mercury has the thinnest atmosphere of all the planets in our solar system and has no way to trap heat near its surface. This means that the night side of Mercury is freezing cold. It is twice as cold as Antarctica, the chilliest place on Earth. The day side of Mercury is even less comfortable. Temperatures there can be twice as hot as a pizza oven! Sunscreen wouldn't help you if you visited there. You wouldn't be able to survive the heat. Since Mercury is closest to the Sun, you might think it is the hottest planet. Surprise! It's not. Venus is!

Does Mercury remind you of our moon? It has many things in common with it. Mercury is tiny—the smallest planet in our solar system. It is only a little larger than the Moon, and just like the Moon, Mercury does not have air. What it does have is lots of craters. Mercury has a rocky crust and its core is made of metals called iron and nickel. The craters are evidence that Mercury has been hit many times by passing space rocks. At the end of this chapter, you will learn how to do an experiment to create craters of your own.

A DEEPER LOOK

The first spacecraft sent to study Mercury was the *Mariner 10*, which flew by Mercury three times in the 1970s. This voyage helped scientists map the dark half of the planet. The spacecraft had to protect itself from the scorching heat on the sunny side of the planet, so it could only take photos of the shadowy side of Mercury, which faces away from the Sun. The *Mariner 10* spacecraft also studied Venus during its journey in space. It was the first mission to photograph two planets in one trip. In 2011 and 2012, NASA sent the *Messenger* spacecraft to study Mercury's surface. It discovered water ice in the craters around Mercury's north pole!

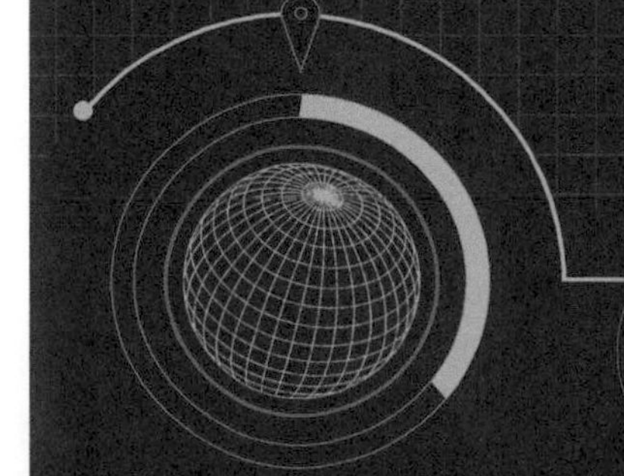

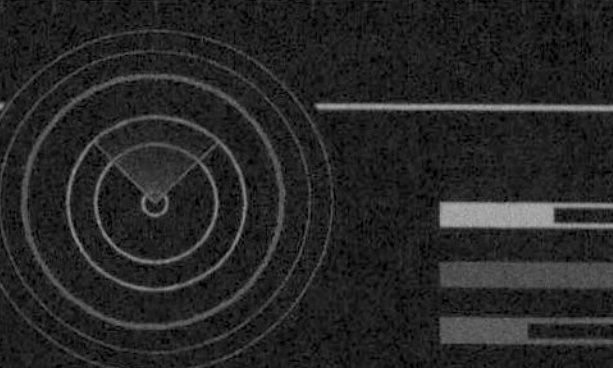

VENUS

DIAMETER:
7,520 MILES

DISTANCE FROM SUN:
67.24 MILLION MILES

DISTANCE FROM EARTH:
23.6 MILLION TO 162.2 MILLION MILES (DEPENDING ON PLACE IN ORBIT)

KNOWN MOONS:
NONE

LENGTH OF 1 DAY:
243 EARTH DAYS

LENGTH OF 1 YEAR:
225 EARTH DAYS

AVERAGE TEMPERATURE:
870 DEGREES FAHRENHEIT

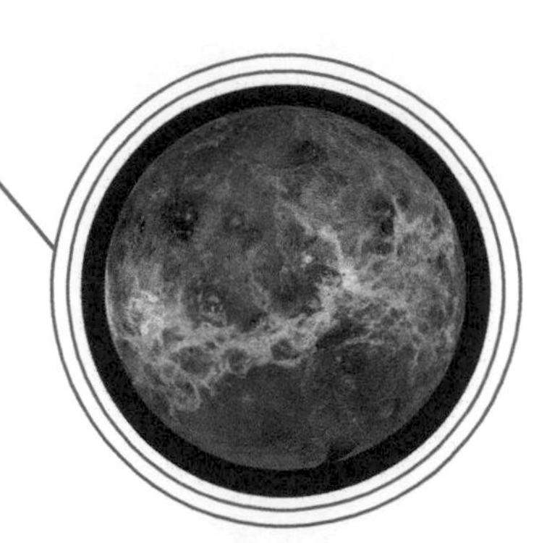

Venus is the second planet from the Sun and is sometimes called Earth's sister planet. Venus is almost as big as Earth. Depending on its location as it orbits the Sun, Venus is also sometimes Earth's closest neighbor. There are tens of thousands of volcanoes on Venus, but we don't know if any are still erupting. The tallest volcano, Maxwell Montes, is taller than Earth's tallest mountain, Mount Everest. Like Mercury, Venus has a core made of iron and nickel. It also has a rocky crust.

Does Venus look like a nice place to visit? It wouldn't be! For one thing, humans would not survive there. Venus's atmosphere is full of poisonous gas. Instead of clouds made of water like Earth's, Venus has clouds made of acid that would burn your skin. These thick clouds do not let much sunlight reach Venus's surface. Although it's not very sunny on Venus, it is the hottest planet in our solar system. The thick atmosphere traps heat, just like glass on a greenhouse.

Most planets rotate from west to east, which is the same direction all the planets orbit the Sun. Venus and Uranus are the only planets that rotate in the opposite direction, east to west. Scientists think a huge asteroid may have hit Venus and slowed its rotation to a standstill, then made it rotate very slowly in the "wrong" direction. It takes 243 Earth days for Venus to complete one rotation on its axis. This is the longest day of any planet. In fact, one Venus day is longer than an entire year on Earth! Venus's orbit is the most circular of any of the planets in our solar system. Other planets' orbits are slightly more oval shaped, or elliptical.

More than 40 spacecraft have explored Venus. In the 1990s, NASA's *Magellan* probe beamed radio waves through Venus's clouds to its surface. This probe helped create pictures of the surface of our slow-spinning sister planet. *Akatsuki* is a Japanese space probe that is orbiting and studying the atmosphere of Venus right now!

Venus is the easiest planet to spot in the sky. It shines brightest right after sunset or just before dawn. This is why it is nicknamed the "evening star" and "morning star."

EARTH

DIAMETER:
8,000 MILES

DISTANCE FROM SUN:
93 MILLION MILES

KNOWN MOONS: 1

LENGTH OF 1 DAY: 24 HOURS

LENGTH OF 1 YEAR:
365 DAYS AND 6 HOURS

AVERAGE TEMPERATURE:
AROUND 60 DEGREES FAHRENHEIT

Welcome home! We live on the third planet from the Sun—Earth. Earth is the fifth largest planet in our solar system and the largest of the terrestrial planets. Scientists nicknamed our planet the "Blue Marble" because it looks like a blue-and-white marble when seen from outer space. Our planet is mostly blue because water covers two-thirds of it. The white shapes are the clouds that swirl around it. Earth may look perfectly round, but it isn't. It bulges out in the middle and is slightly flat on its top and bottom. This shape is called an **oblate spheroid**. Earth's rotation caused it to be shaped this way.

Earth feels solid beneath our feet, but deep below the rocky crust, there is very hot liquid rock and metal. Earth's core is a solid ball of iron and nickel, but it is surrounded with liquid iron and nickel. The heat inside Earth's core can cause earthquakes and volcanoes on the surface.

Earth has four seasons. Why? Earth is slightly tilted. Its axis is not pointing straight up and down, but it leans just a little. This means that the north pole leans toward the Sun for part of the year. During this time, the northern part of the planet has longer, warmer days with more direct sunlight. This is summer for people in the northern hemisphere. The south pole leans away from the Sun during this time, so it is winter for people who live on the southern part of the planet. When Earth is halfway around its orbit six months later, the seasons switch.

Earth's atmosphere is like a thick blanket. It has five layers. The two layers closest to the planet are called the troposphere and the stratosphere. The troposphere is about seven miles thick. This is where Earth's weather happens. The troposphere also contains the air we breathe. The stratosphere is on top of the troposphere. It reaches about 30 miles above Earth. The stratosphere and the layers above it also protect us from falling space objects, like meteoroids.

WHAT MAKES US DIFFERENT?

Think of all of the things you can do on Earth. You can breathe the air, climb a mountain, and swim in the ocean. Earth is the only place we know of in our solar system where you can do those things. Earth has perfect conditions to make animal and plant life survive.

Other planets do not have the right conditions for life. If our planet were farther away from the Sun, it would be too cold for plant and animal life. If it were closer, it would be too hot. Not every planet has a livable atmosphere. Some have an atmosphere too poisonous to breathe. Gas planets, like Saturn or Jupiter, don't even have a solid surface that you can walk on!

The water on our planet also makes life possible. Earth is the only place that has water in all its forms: gas, solid, and liquid. Earth is home to more than 8.7 million species. It is truly a special place!

EARTH'S MOON

DIAMETER:
2,155 MILES

DISTANCE FROM EARTH:
238,855 MILES

LENGTH OF ORBIT AROUND EARTH:
27.3 EARTH DAYS

AGE:
4.5 BILLION YEARS

AVERAGE TEMPERATURE:
ABOUT 270 DEGREES FAHRENHEIT

Are you ready to leave Earth? The Moon is our next destination! People have sent thousands of satellites into Earth's orbit, but the Moon is our only natural satellite. It travels around our planet in a path shaped like a flattened circle, or ellipse. It takes the Moon 27 days to orbit Earth and 27 days to rotate on its axis. This means we always see the same side of the Moon from Earth.

Our Moon is one-quarter of the size of Earth. At night, it seems to glow, but it doesn't actually shine like a star. The light you see is reflected from the Sun. The Moon has no atmosphere to protect its surface from incoming objects. If you look through binoculars or a telescope, you will see that the Moon's surface is covered with craters. These craters were caused by meteorites and asteroids crashing into the Moon millions of years ago.

What would it be like to visit the Moon? For one thing, you could jump six times higher than you could on Earth! This is because the Moon is less massive than Earth, so it has less gravity. You would have to wear a spacesuit since there is no air to breathe. Earth's Moon is the only place beyond our planet that astronauts have visited. Because there is no wind or rain, the footprints of those astronauts will stay on the surface of the Moon for billions of years!

The Moon affects the Earth in an important way. The Moon may be small, but it is close to our planet. The Moon's gravity pulls more on the part of the Earth closest to it and less on the part of the Earth farther from it. As the Earth spins, this difference causes high and low ocean tides. The Sun also contributes a little bit. The Sun has more gravity, but it is much farther away than the Moon. Other planets and moons experience tides, too. Jupiter causes strong enough tides on its moons to fuel volcanoes on one of them!

The Moon's shape in the sky seems to change as it travels around Earth every 27 days. This is because the Sun is always shining on exactly half of the Earth and half of the Moon, but their relative positions in space are changing. When the Moon looks like a crescent, we are seeing a small slice of the lit side. When the Moon is full, we are seeing the entire lit side. The shape of an almost-full moon is called a gibbous. Did you know you can predict what time it is based on the phase of the Moon and where it is in the sky? For example, if you see a full moon rising, it must be around sunset. Another fun fact is that when astronauts went to the Moon, they saw the Earth go through phases, too.

LUNAR ECLIPSES

A **lunar eclipse** is when the Earth, Moon, and Sun all line up. On these special nights, you can watch the full moon slowly pass through the darkest part of Earth's shadow. The Moon is gradually darkened by the shadow until

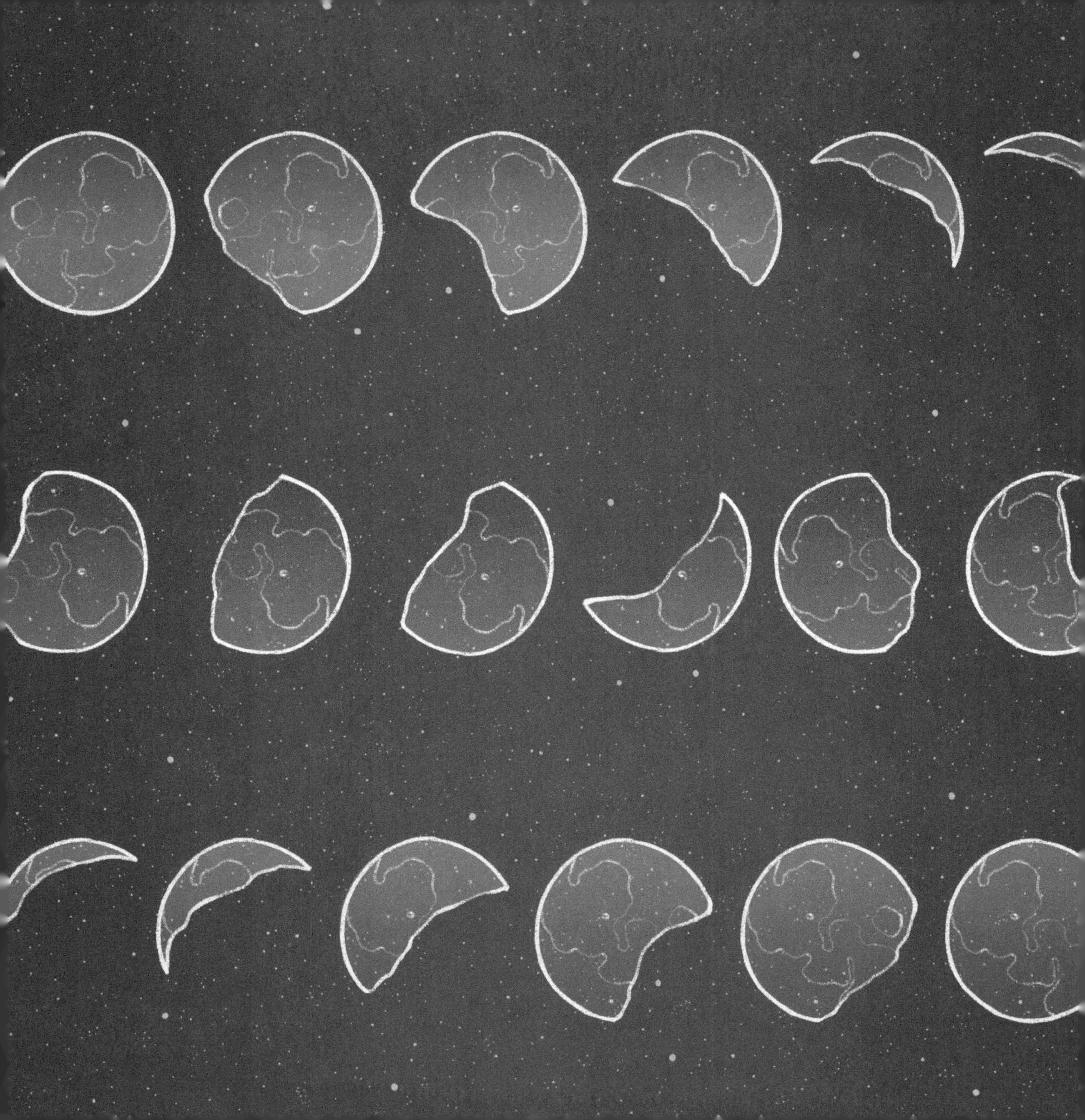

it is on the opposite side of our planet from the Sun. When everything is lined up, the Moon looks very dim and sometimes can appear a dark red color. As the Moon continues its orbit, it gradually comes out of Earth's shadow and becomes a full moon once again.

MOON LANDINGS

Moon landings are times when scientists and engineers have worked together to put spacecraft or people on the Moon. Astronauts have been sent to the Moon six times. Twelve astronauts from those moon landings actually walked on the Moon! There have also been many landings without people aboard. These are called robotic missions.

TAKE A LOOK!

The illustration shows the stages of a lunar eclipse. The parts of the Moon that are missing show where Earth's shadow blocks it.

▶ LUNA 2 & LUNA 3

Luna 2 was the very first spacecraft to land on the Moon. It was launched in 1959 by the Soviet Union—present-day Russia. Unfortunately, *Luna 2* crash-landed into the Moon just 33.5 hours after it was launched. Shortly thereafter, the Soviet Union launched *Luna 3*, which took the first pictures of the far side of the Moon.

▶ APOLLO 8

Apollo 8 was the first spacecraft to carry people to the Moon. The crew spent a day orbiting the Moon and sent back the first photographs of Earth from space. *Apollo 8* left the Moon's orbit and headed home on Christmas Day, December 25, 1968.

▶ APOLLO 11

In 1969, the United States became the first country to land humans on the Moon. *Apollo 11* took them there. When astronaut Neil Armstrong stepped on the Moon, he said, "One small step for man, one giant leap for mankind."

▶ CHANG'E-4

In January 2019, this robotic Chinese spacecraft was the first to successfully land on the far side of the Moon. It is part of the Chinese Lunar Exploration Program, which is a series of robotic Moon missions. The spacecraft includes a rover called *Yutu-2*. Studying this unexplored area helps scientists learn more about the history of Earth and the solar system. *Chang'e-4* continues to explore the Moon today.

A DEEPER LOOK

NASA has sent four robotic vehicles, called rovers, to Mars to study the planet. These four rovers are named *Sojourner*, *Spirit*, *Opportunity*, and *Curiosity*.

Rovers land on the planet's surface and use special wheels to drive around and gather information. Scientists on Earth send commands to tell the rovers where to go. Rovers help scientists do experiments to understand what planets are made of without having to send people there. Special instruments allow scientists to study the materials in different types of rocks. The Mars rovers found evidence that there used to be water on Mars!

The next Mars rover, named *Perseverance*, is scheduled to launch in July 2020 and land on Mars in February 2021. Guess who named it? A seventh-grade boy who won a nationwide contest! *Perseverance* will stay on the planet for at least one Mars year (687 Earth days). It will search for signs of past microscopic (very small) life, study rocks, report on the weather, and collect samples to bring back to Earth. It will also study whether humans can explore Mars themselves one day.

MARS

DIAMETER:
4,222 MILES

DISTANCE FROM SUN:
142 MILLION MILES

DISTANCE FROM EARTH:
33.9 MILLION TO 140 MILLION MILES
(DEPENDING ON PLACE IN ORBIT)

KNOWN MOONS:
2 (PHOBOS AND DEIMOS)

LENGTH OF 1 DAY:
24 HOURS AND 40 MINUTES

LENGTH OF 1 YEAR:
687 EARTH DAYS

AVERAGE TEMPERATURE:
-80 DEGREES FAHRENHEIT

Do you ever dream of visiting Mars? It is the planet most like Earth, but you would need to wear a spacesuit to safely visit, and it would take about seven months to get there. A day on Mars is just barely longer than our 24-hour day. It would be difficult to live on Mars because there is no oxygen to breathe, and it is very cold.

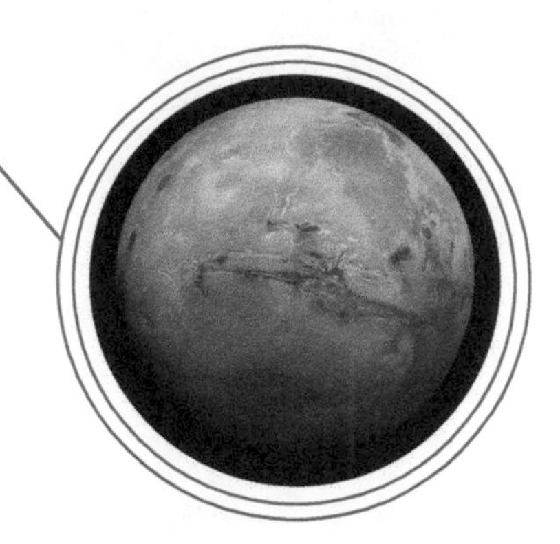

Billions of years ago, Mars was almost as warm as Earth. These days, Mars is farther from the Sun. It gets much colder than the coldest places on Earth. If you visited, you'd see not one but two little potato-shaped moons in the sky: Phobos and Deimos.

Mars is the fourth planet from the Sun and the second smallest planet in our solar system. Mars is nicknamed the "Red Planet." Can you guess why? The soil on Mars is full of iron oxide, or rust. This makes the whole planet look red. Since there is no rain, Mars is always dry and dusty. Just like Earth, Mars has mountains, ice on its north and south poles, sand dunes, and deserts. Mars also has volcanoes. Olympus Mons is a giant volcano that is three times as high as Mount Everest, the tallest mountain on Earth. A huge canyon called Valles Marineris looks like a giant cut in the planet's side.

In 2017, researchers learned that a giant asteroid crashed into Mars about three billion years ago. It may have landed in one of Mars's largest ancient oceans. causing tsunamis, or giant waves. The push and pull of the waves left behind a large crater. Scientists believe the crater will prove Mars had lakes, rivers, and oceans, just like Earth. In 1996, scientists found something that looked like fossils of bacteria in a meteorite on Earth that came from Mars. This suggested that Mars may have once had living things. Other scientists disagree that life existed there.

There have been dozens of missions to Mars since 1960. Some spacecraft just fly past. Others orbit the planet for years, sending back photographs to help us understand the planet better. In November 2018, NASA's *InSight* spacecraft landed safely on Mars's surface. It is studying what is under the planet's crust and has detected "Marsquakes."

THE ASTEROID BELT

When our solar system was formed, there were some rocky "leftovers" that never became planets or moons. These rocky pieces are called **asteroids**. Like planets, asteroids orbit the Sun. They can be as small as boulders or hundreds of miles across. Most of them are small and orbit the Sun far from Earth, in between Mars and Jupiter, in a region called the asteroid belt.

Sometimes asteroids have unusual orbits and crash into Earth! Most of the time, Earth's atmosphere protects us and the asteroid burns up as a meteor. This is one reason it's hard to find craters on Earth but easy to find them on planets and moons with no atmosphere. Larger asteroids with unusual orbits are rare but can be dangerous. Many have crashed into Earth in the past and will again in the future. In fact, scientists learned dinosaurs went extinct shortly after a huge asteroid slammed into the planet millions of years ago.

When an astronomer discovers an asteroid, sometimes he or she gets to name it. Many asteroids have names from mythology. Some are named after famous astronomers. One asteroid was even named after an astronomer's cat who was also named after a TV science-fiction character, Mr. Spock!

Most asteroids are made of rocks and stone. There are three main kinds: C-type (carbonaceous, or "clay"), S-type (silicaceous, or "stony"), and M-type (metallic). Even though there are millions of asteroids in the asteroid belt, they are so far apart that you could fly through the belt and never see an asteroid through your window. Read on to find out what the largest object in the asteroid belt is!

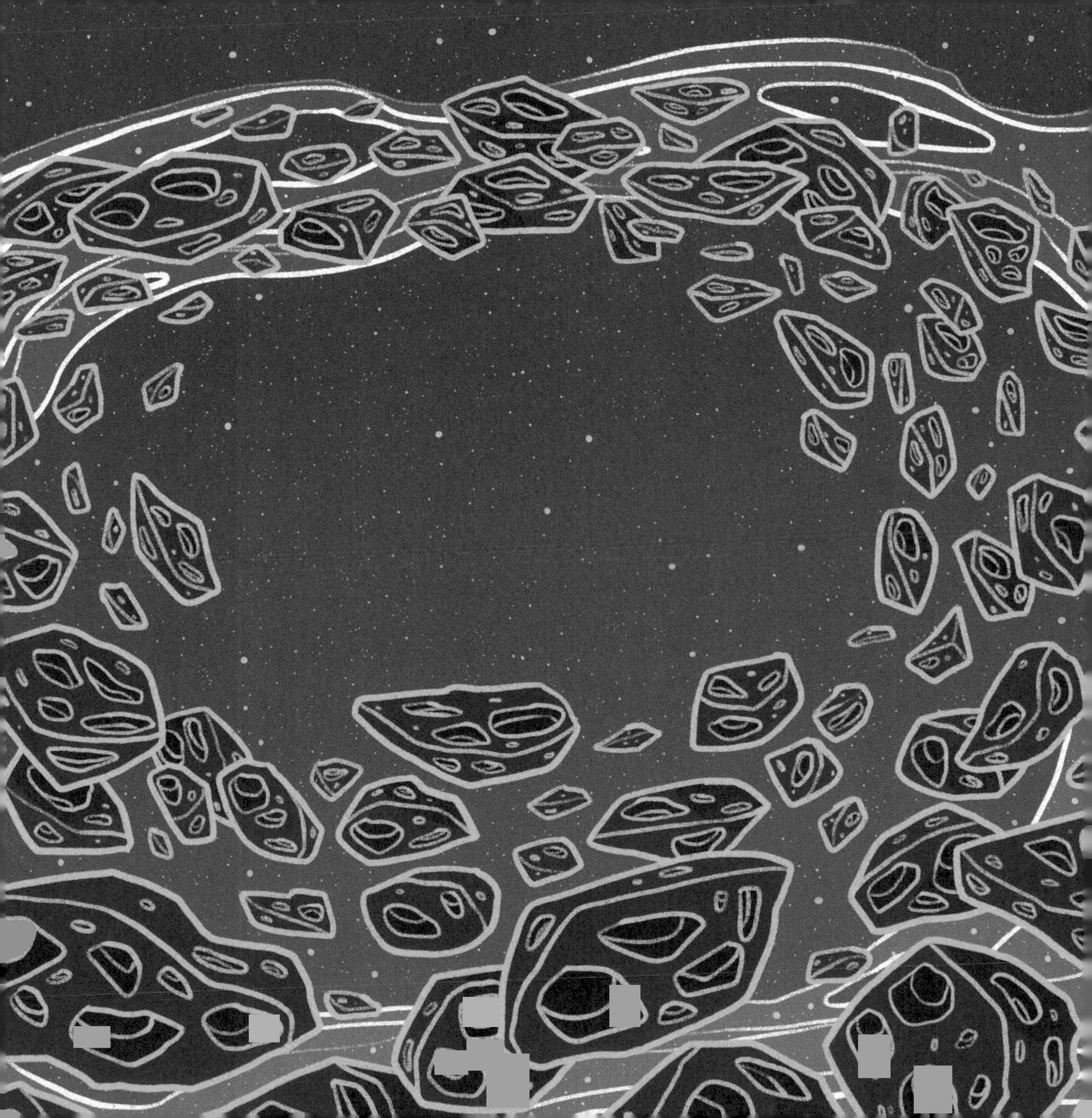

CERES

DIAMETER:
600 MILES

DISTANCE FROM SUN:
257 MILLION MILES

DISTANCE FROM EARTH:
344 MILLION MILES

KNOWN MOONS:
NONE

LENGTH OF 1 DAY:
9 HOURS

LENGTH OF 1 YEAR:
1,680 EARTH DAYS

AVERAGE TEMPERATURE:
-100 TO -225 DEGREES FAHRENHEIT

Something much bigger than the average asteroid cruises through the asteroid belt. Its name is Ceres and it is the closest dwarf planet to Earth. Ceres is by far the largest object in the asteroid belt. It contains about one quarter of all the material in the belt. Despite its large size, it is still much smaller than our Moon. The diameter is about the same as the width of the state of Texas.

Ceres has a thin, dusty, and rocky surface that covers an icy **mantle** and a rocky or metallic inner core. Scientists think the interior of Ceres contains a large amount of salty ice and water. Ceres's north and south poles also probably have ice. Because of its location in the asteroid belt, Ceres takes about 4.6 Earth years to make one orbit around the Sun. Unlike Earth, Ceres doesn't have much of an atmosphere, but scientists think it may have small amounts of water vapor, the gas form of water.

Ceres was discovered by an Italian astronomer named Giuseppe Piazzi in 1801. He was looking for a star but found Ceres instead! It was the easiest object in the asteroid belt to see from Earth. Ceres was called a planet for almost 50 years. Later, when other objects were found in Ceres's orbit, astronomers understood that it wasn't a true planet after all. That's when scientists started calling Ceres an asteroid. One hundred and fifty years later, in 2006, scientists classified Ceres as a dwarf planet because it was much bigger and different from its neighbors. We will learn about more dwarf planets in chapter 5.

Ceres is the first dwarf planet to be visited by spacecraft. In 2015, the spacecraft *Dawn* orbited Ceres to study its surface and history. *Dawn* showed us that dwarf planets may have had oceans at one time—and possibly still could. Ceres is a place that scientists are excited to study because many of the building blocks for life are there. These include certain molecules containing the element carbon as well as plenty of water and ice.

EXPLORE MORE!

Make your own crater model! All you need is a handful of marbles, three cups of flour, and a baking tray. Sprinkle the flour into an even layer on the baking tray and make the surface smooth. Drop the marbles onto the flour and carefully remove them. You will see saucer-shaped craters! Try experimenting to see if the craters are bigger or smaller when you drop the marbles from higher (more impact energy) or lower (less impact energy).

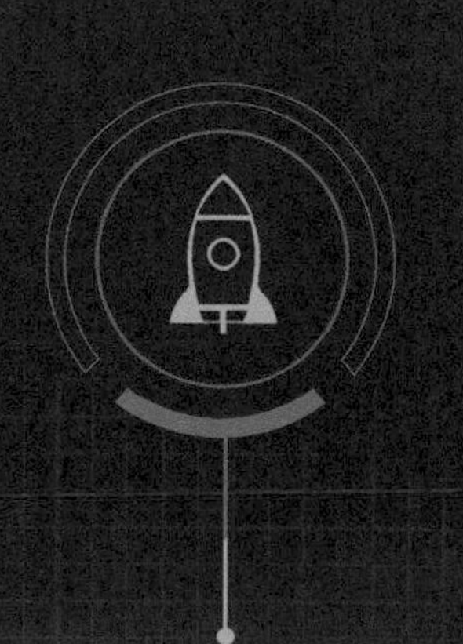

Chapter Four

GAS GIANTS AND ICE GIANTS

Imagine a planet that has no surface to walk on! Our solar system has four planets of this type. They are the biggest planets in our solar system. Two of them are called gas giants: Jupiter and Saturn. Two of them are called ice giants: Uranus and Neptune. We'll learn more about ice giants later.

Gas giants have very thick atmospheres. When we look at these planets, we see the top of their clouds, not the ground. These planets are mostly gas and liquid, but they may have small cores. Jupiter and Saturn are so large that their small cores are actually bigger than all of planet Earth!

JUPITER

DIAMETER:
88,846 MILES

DISTANCE FROM SUN:
484 MILLION MILES

DISTANCE FROM EARTH:
483 MILLION MILES

KNOWN MOONS:
79

LENGTH OF 1 DAY:
LESS THAN 10 HOURS

LENGTH OF 1 YEAR:
11.8 EARTH YEARS

AVERAGE TEMPERATURE:
-240 DEGREES FAHRENHEIT

It's time to visit the biggest planet in our solar system: Jupiter, the fifth planet from the Sun. Jupiter is enormous. If you combined all of the other planets together into one, Jupiter would still be twice as big. Earth is tiny compared with Jupiter. You could fit 11 planet

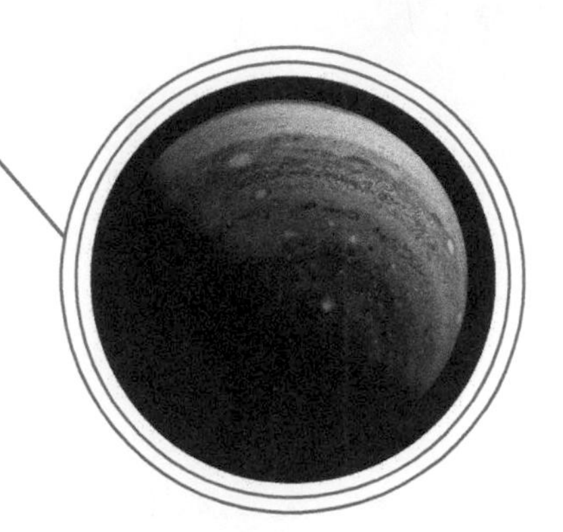

Earths side by side across Jupiter's equator. Despite this fact, Jupiter is still much smaller than the Sun.

Jupiter is a huge ball of hydrogen and helium gases. Scientists believe it is quite stinky because its atmosphere has a lot of ammonia in it.

Even though Jupiter is massive, it spins faster than any other planet. One day on Jupiter is less than 10 hours. The planet's intense spinning causes giant windy storms. Some of the winds gust up to 300 miles per hour! These storms can last a very long time. Jupiter's Great Red Spot is a storm that has been raging for more than 300 years. The Great Red Spot is larger than Earth and swirls above the rest of Jupiter's clouds. Jupiter's winds blow the clouds into colorful bands that look like stripes around the planet.

Jupiter has lots of company—79 known moons orbit the giant. Ganymede is the largest moon in the solar system. It is bigger than the planet Mercury and only a little smaller than Mars. Ganymede would be called a planet if it was orbiting the Sun instead of Jupiter.

Astronomers have always been curious about Jupiter. In 1995, the *Galileo* spacecraft began orbiting the planet, taking photographs of Jupiter's storms, moons, and dusty rings (which are too faint to be seen from Earth). In 2016, the *Juno* space probe arrived at Jupiter and began orbiting to study the planet, its moons, and its environment. It will stay there until 2021. *Juno* has recorded new information about Jupiter's temperature, size, and weather. It has also taken amazing photographs.

A DEEPER LOOK

Many planets have a magnetic field. What is a magnetic field? It is an invisible force that can either bring objects closer together or push them apart. When a magnetic field surrounds a planet, it's called a magnetosphere. The gas and ice giants have very strong magnetic fields. If you could see Jupiter's magnetosphere, it would be 20 times bigger than our Sun! Jupiter's magnetic field is so strong that high energy particles from the Sun, called solar wind, bounce off it. Earth's magnetic field is very strong, too. When solar wind interacts with Earth's atmosphere and magnetic field, it can cause a beautiful light show in the night sky known as an aurora. These auroras are sometimes called the northern lights or southern lights.

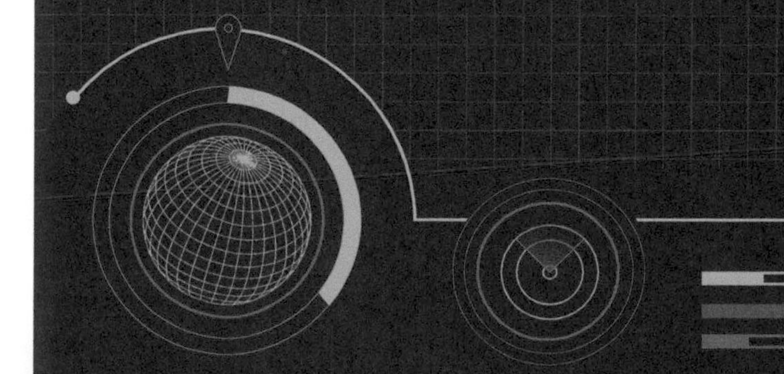

SATURN

DIAMETER:
72,367 MILES

DISTANCE FROM SUN:
886 MILLION MILES

DISTANCE FROM EARTH:
746 MILLION TO 1 BILLION MILES (DEPENDING ON PLACE IN ORBIT)

KNOWN MOONS:
53

LENGTH OF 1 DAY:
10 HOURS AND 40 MINUTES

LENGTH OF 1 YEAR:
29.4 EARTH YEARS

AVERAGE TEMPERATURE:
-210 DEGREES FAHRENHEIT

Do you like things with rings? Then you'll love Saturn! This gas giant is the sixth planet from the Sun and the second largest planet in our solar system. It's also the most distant planet that you can see without a telescope from Earth, and it's the only planet that is less dense than water. This means that if you could fit Saturn into your bathtub, it would float! Like Jupiter, Saturn is a giant planet made up of hydrogen and helium gases.

Saturn is famous for its nine rings. Even though the rings look like they're solid, they're actually made up of billions of chunks of ice, dust, and rock. Some of these particles are as small as a grain of sand, whereas others are as large as cars. The rings seem to shine because the ice particles reflect light. Galileo was confused when he looked at Saturn through his telescope. He thought he might be seeing three planets clumped together—or maybe a planet with handles like a mug. It wasn't until we had better telescopes that astronomers could see the rings clearly. You can easily see Saturn's rings with even the smallest telescope.

Saturn is super windy. Sometimes the wind blows up to 1,000 miles per hour. That is 10 times faster than a hurricane on planet Earth. Saturn's atmosphere is made up of different gases, and some do not mix. When Saturn spins, high winds blow the gases around the surface of the planet, forming stripes. Each of the stripes is a different mixture of gas. Saturn's stripes and clouds are less colorful than Jupiter's. This might be because Saturn is farther from the Sun, so it receives less energy and heat.

Saturn's largest moon is Titan, which is the second largest moon in the solar system and the only moon with a thick atmosphere. Titan has thick orange clouds and is almost twice the size of Earth's Moon. Titan has large lakes made of methane. This substance is a gas on Earth, but a liquid on Titan because it is so cold there.

The *Cassini* spacecraft studied Saturn from orbit for 13 years. It learned many secrets about Saturn's rings and moons. Then scientists decided to send *Cassini* into Saturn's clouds to learn more about the planet's atmosphere. *Cassini* dove into Saturn in September 2017. In doing so, the spacecraft burned up, but not before sending scientists a report of what it had learned. *Cassini* also sent a small probe called Huygens to land on Titan. The probe sent back lots of information about Titan's thick atmosphere, and survived for 90 minutes on the surface, long enough to take a picture!

URANUS

DIAMETER:
31,584 MILES

DISTANCE FROM SUN:
1.79 BILLION MILES

DISTANCE FROM EARTH:
1.6 BILLION TO 1.9 BILLION MILES (DEPENDING ON PLACE IN ORBIT)

KNOWN MOONS:
27

LENGTH OF 1 DAY:
17 HOURS AND 14 MINUTES

LENGTH OF 1 YEAR:
84 EARTH YEARS

AVERAGE TEMPERATURE:
-353 DEGREES FAHRENHEIT

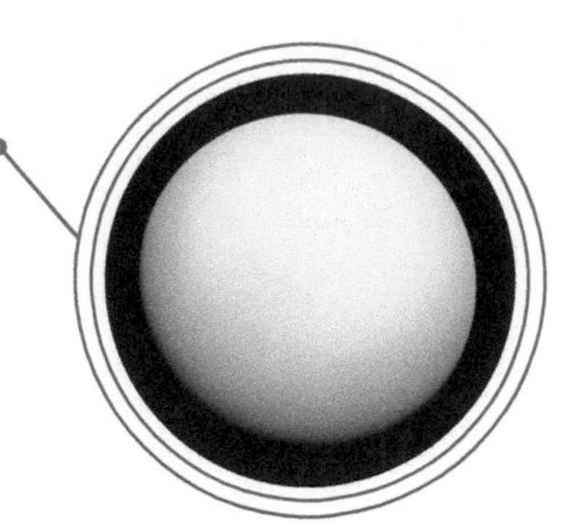

We have now traveled more than 1.5 billion miles into space! We've arrived at Uranus, the seventh planet from the Sun. It's four times larger than Earth but about half as big as Saturn.

Uranus is called an ice giant. This is because it's made mostly of a slushy mix of water, ammonia, and a gas called methane. This slush floats around a small rocky core. The slush is very cold, but the core is very hot—9,000 degrees Fahrenheit! Uranus's atmosphere contains mostly helium and hydrogen, with some methane mixed in. It's the methane that makes Uranus look blue.

Uranus has 13 known rings around it. The rings on the outside are bright, but the ones closer to the planet are dark. Do you notice anything else that makes Uranus's rings different from Saturn's? Instead of looking like a Hula-Hoop around the planet, Uranus's rings look more like a Hula-Hoop standing on end. This is because Uranus spins on its side. Scientists think an asteroid crashed into it millions of years ago and knocked the planet onto its side.

It takes Uranus 84 Earth years to complete one orbit. This means its summer and winter seasons last 21 years each. Spring and fall last 42 years each. Uranus is the only other planet besides Venus that rotates clockwise. A massive asteroid may have caused its backward rotation, too.

Uranus has 27 known moons. Miranda is the smallest. Ariel is the brightest. Umbriel is the darkest. Oberon is covered in craters. Titania is the largest.

Only one spacecraft has visited Uranus. NASA's *Voyager 2* made the first and, to date, only visit to the distant planet in 1986. Nine years after it launched, *Voyager 2* reached Uranus. Once there, it spent only about six hours gathering information about Uranus's rings and moons.

A DEEPER LOOK

Some astronomers think there is an undiscovered planet on the edge of our solar system. They call it Planet X or Planet Nine. If there is such a planet, it would help explain the strange orbits of some smaller objects in the Kuiper Belt. The Kuiper Belt (more on this in chapter 5) is an area filled with icy objects beyond Neptune. Scientists will soon explore the Belt with strong telescopes to see if they can find the mystery planet. Who knows? Maybe they will discover a new planet in our solar system during your lifetime!

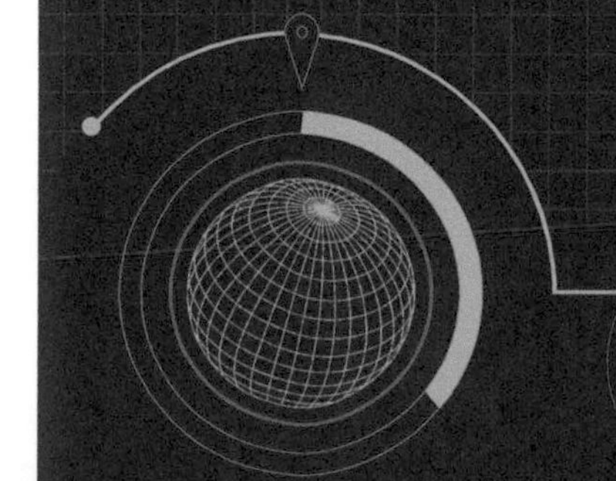

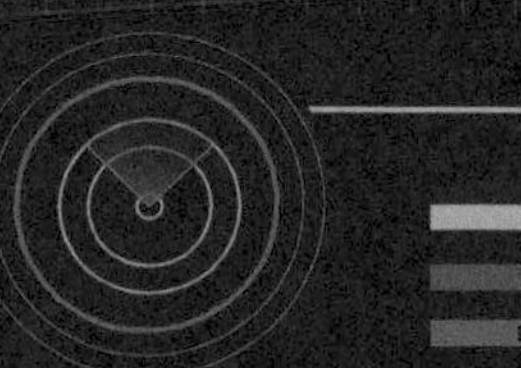

NEPTUNE

DIAMETER:
30,598 MILES

DISTANCE FROM SUN:
2.8 BILLION MILES

DISTANCE FROM EARTH:
2.7 BILLION TO 2.9 BILLION MILES (DEPENDING ON PLACE IN ORBIT)

KNOWN MOONS:
14

LENGTH OF 1 DAY:
16 HOURS AND 6 MINUTES

LENGTH OF 1 YEAR:
165 EARTH YEARS

AVERAGE TEMPERATURE:
-392 DEGREES FAHRENHEIT

We've made it to the eighth and last planet in our solar system: Neptune. Even though it's the smallest of the gas and ice giants, Neptune is still four times the size of Earth. Neptune is very far away from the Sun—30 times farther than Earth. This makes it an extremely cold and very dark place. Neptune is so far away that it's the only planet that you can't see without a telescope in the night sky. Neptune has an oval-shaped orbit. For 20 years, it was farther away from the Sun than Pluto! Neptune's orbit is so large that it hasn't even gone around the Sun once since it was discovered.

In many ways, Uranus and Neptune are like twins. Like Uranus, Neptune has a group of thin rings— at least five of them. The planet also looks blue like its neighbor because its atmosphere contains methane gas. Neptune also has an Earth-shaped rocky core covered in a slushy mix of water and ammonia. Some scientists think Neptune could be hiding a giant ocean of very hot water under its thick clouds! Neptune has 14 known moons. Its largest moon is named Triton. Triton is the only known moon in our solar system that orbits its planet in the opposite direction the planet is rotating.

Besides being cold and dark, Neptune is very windy. It has some of the fastest winds in the solar system. These winds cause strong storms. The Great Dark Spot was a giant swirling storm that scientists found on Neptune in 1989. *Voyager 2* discovered and photographed it. Five years later, scientists used the Hubble Space Telescope to look at Neptune. The Great Dark Spot was gone.

In 1612, Galileo saw Neptune in his telescope and thought it was a star. In 1846, astronomers realized that something large was making Uranus orbit in a strange way. They used math and careful observations to discover that this "star" was actually a planet—it was just the right size and in the right location to explain Uranus's odd orbit. NASA's *Voyager 2* is the only spacecraft that has studied the distant planet up close.

EXPLORE MORE!

You can see Jupiter's Galilean Moons from your own backyard! On a dark night, ask a grown-up to help locate Jupiter. Use a pair of binoculars to find four tiny points of light near Jupiter. They look like stars, but they are moons! Draw a picture of what you see. The moons are always on the move. If you look for them on a different night, they will be in different positions! If you see only three, one of them may be in front of or behind Jupiter.

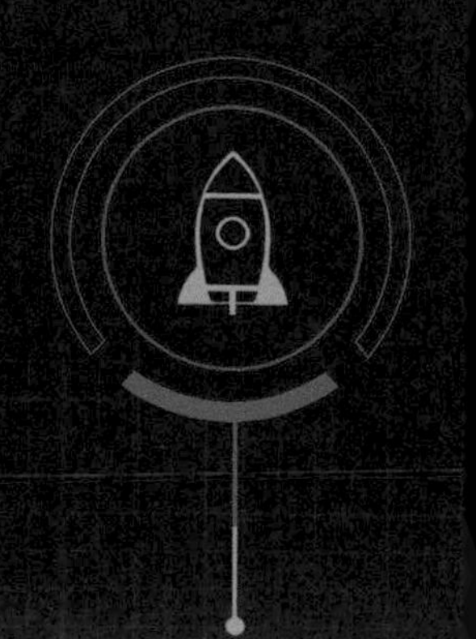

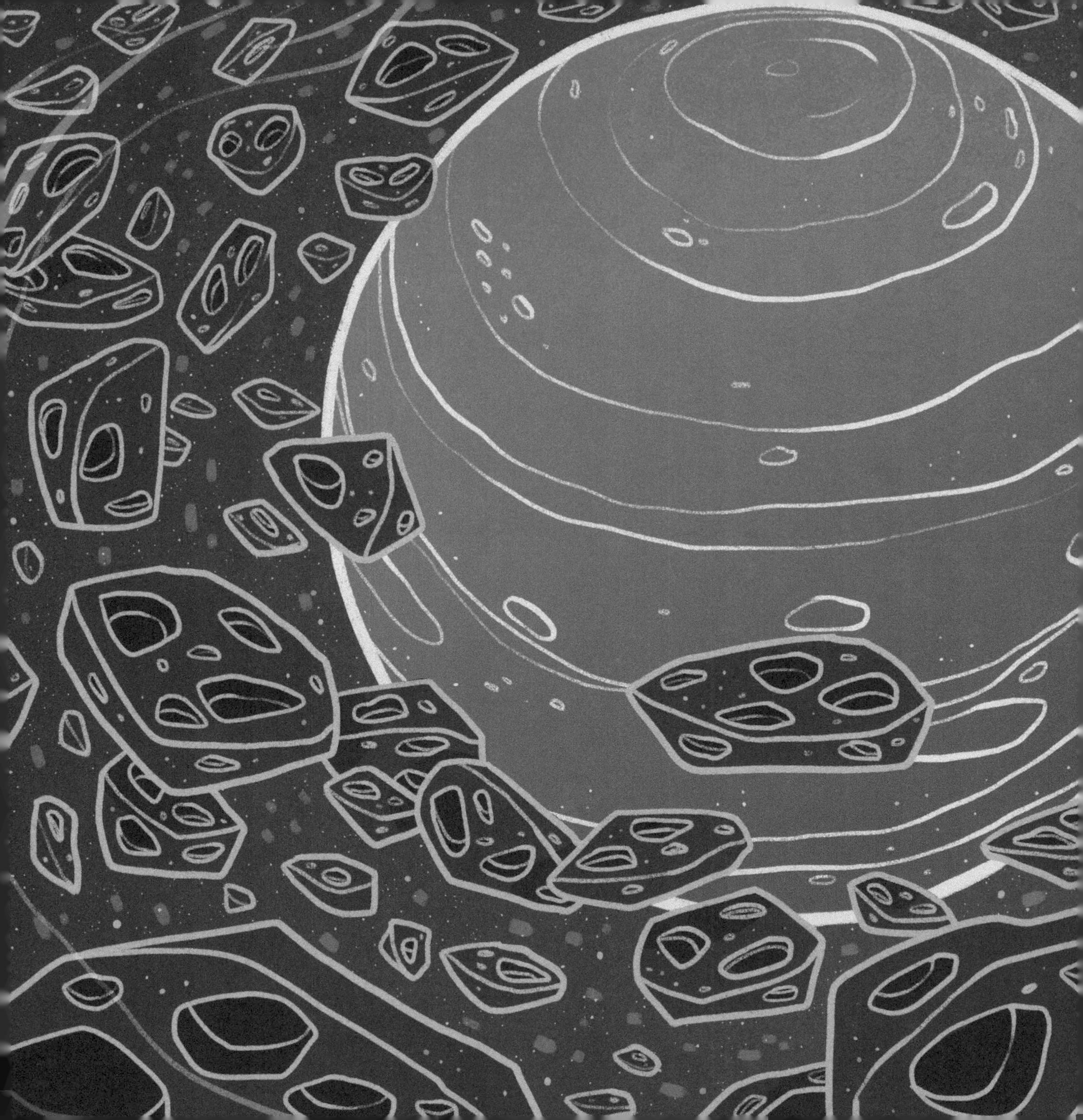

Chapter Five

THE KUIPER BELT AND DWARF PLANETS

Beyond Neptune, the Kuiper Belt sits like a giant doughnut in space. Look closely and you'll find that the doughnut is made of millions of small objects—rocks, ice chunks, comets, and dwarf planets. Four out of the five dwarf planets in our solar system orbit the Sun in the Kuiper Belt. Pluto is the most famous Kuiper Belt object.

PLUTO

DIAMETER:
1,470 MILES

DISTANCE FROM SUN:
4.5 BILLION MILES

DISTANCE FROM EARTH:
2.6 BILLION TO 4.6 BILLION MILES (DEPENDING ON PLACE IN ORBIT)

KNOWN MOONS: 5

LENGTH OF 1 DAY:
6 EARTH DAYS

LENGTH OF 1 YEAR:
248 EARTH YEARS

AVERAGE TEMPERATURE:
-387 DEGREES FAHRENHEIT

Traveling to dwarf planet Pluto would take a very, very long time. It is so far away from the Sun that it has a huge orbit. It must travel 22.6 billion miles to circle the Sun one time. It's no wonder it takes 248 Earth years to do it! Pluto has an orbit shaped much more like an oval than a circle, which means Pluto's distance from the Sun changes as it orbits. Pluto rotates much slower than Earth. It takes Pluto 153 hours to complete one rotation—more than six Earth days! Like Venus and Uranus, Pluto rotates backward, or in a clockwise direction.

Tiny Pluto is much smaller than Earth—it's only half as wide as the United States! But this dwarf planet is a very interesting place. It has many things you'd see on Earth: valleys, mountains, and plains. It even has a glacier—which is a giant sheet of ice—that is shaped like a heart! Pluto has a rocky core that is covered by water ice. Pluto's hard outer crust is made of frozen methane and nitrogen gases. The planet has five moons. Charon is the largest.

In fact, it's more than half the size of Pluto, which means Pluto and Charon actually orbit each other.

Pluto's atmosphere changes depending on where it is in its orbit. When Pluto is closer to the Sun, ice on its surface warms up and becomes gas. The gases form a thin atmosphere. As the dwarf planet moves farther away from the Sun, the gases freeze again and fall back onto the surface as red snow.

Pluto was discovered in 1930 by a young American astronomer named Clyde Tombaugh. Astronomers had been looking for Planet X beyond Neptune but could not prove it existed. Tombaugh spent a lot of time looking for Planet X. Finally, he found a moving object—it was Pluto!

In 2015, NASA's *New Horizons* spacecraft was the first to explore the dwarf planet and its moons up close. The spacecraft took incredible

A DEEPER LOOK

In 1930, Clyde Tombaugh used a device called a "blink comparator" to discover Pluto. It was a special microscope that scientists used to look for changes in a space object's position by comparing two photographs. Tombaugh noticed a moving object in a set of photos taken six days apart on January 23 and January 29, 1930. The object moved too slowly to be an asteroid. He introduced Pluto, the newest member in our family of planets. It would be our ninth planet for almost 70 years.

A DEEPER LOOK

The *New Horizons* probe is an important part of space history. It is the fastest human-made object ever launched from Earth. *New Horizons* zoomed toward Pluto at 36,373 miles per hour. If a plane could travel that fast, it could fly around the world in about 40 minutes! *New Horizons* is still exploring space. Who knows what it will find next? In the meantime, the probe's data will tell us more about dwarf planets and the Kuiper Belt than we have ever known.

photographs that showed Pluto's amazing features. Pluto probably has a water-ice ocean layer beneath its surface!

PLUTO'S PAST AS A PLANET

Pluto was once the ninth planet in our solar system. What happened? As astronomers found other Pluto-like objects in the outer solar system, they began to wonder if tiny Pluto could really be considered a planet. Scientists spent years discussing whether other dwarf planets should be called planets, or if Pluto should be named as a dwarf planet.

The International Astronomical Union, which is a group of experienced astronomers, decided in 2006 that Pluto should be called a dwarf planet. They explained that Pluto did not meet the three rules used to define a full-size planet. Although Pluto follows two of the rules—it orbits the Sun and is able to keep a round shape—Pluto does not have a magnetic field strong enough to clear other objects from its orbit. This is why scientists reclassified it as a dwarf planet.

Astronomers and space fans have spent years discussing and disagreeing about Pluto's status. But no matter what we call it, Pluto is an amazing world that can tell us a lot about how the solar system formed and evolved. Let's learn about the three other dwarf planets that keep Pluto company in the Kuiper Belt!

A DEEPER LOOK

What is it like to be an astronaut? Things are very different in space. One thing you'd notice right away is that you and everything inside the spacecraft feels weightless. Everything floats around because there is no gravity holding you to the surface of Earth. Can you imagine grabbing your pencil as it floats through the air? How about having to use a wet towel to bathe since water would not stay in a bathtub? To sleep, you would need to be strapped into your bed! Things are definitely weird in space, but most astronauts agree that space travel is a beautiful and exciting experience!

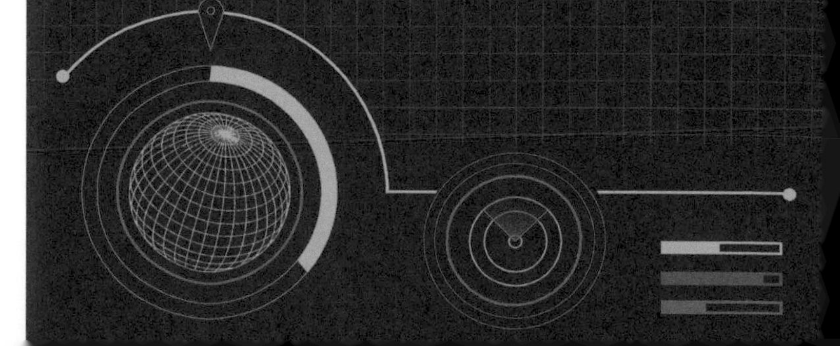

HAUMEA

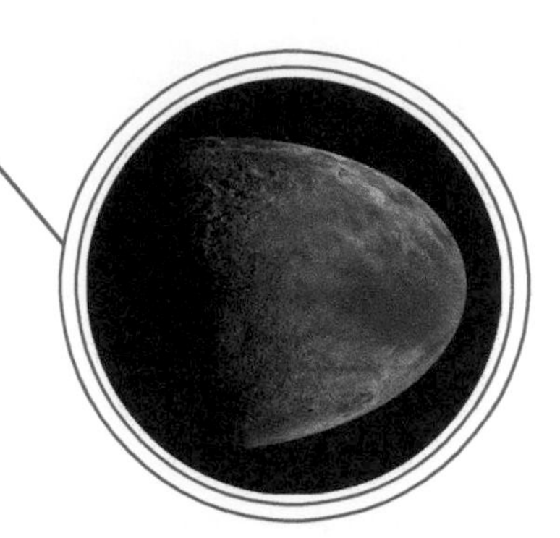

DIAMETER:
1,014 MILES

DISTANCE FROM SUN:
4.7 BILLION MILES

DISTANCE FROM EARTH:
4 BILLION MILES
(AVERAGE CLOSEST DISTANCE)

KNOWN MOONS: 2

LENGTH OF 1 DAY:
3.9 HOURS

LENGTH OF 1 YEAR:
284 EARTH YEARS

AVERAGE TEMPERATURE:
-402 DEGREES FAHRENHEIT

Our next stop is Haumea, an oddly shaped little dwarf planet that is almost the same size as Pluto. This dwarf planet was named after the Hawaiian goddess of childbirth.

Haumea's shape reminds some people of a football on its side. It's twice as wide as it is "tall." Scientists think Haumea has this shape because it spins so fast. It rotates once every four hours! Almost nothing in our solar system spins as quickly. For every one Earth day, Haumea has six! When Haumea is rotating, it looks a little like an egg spinning on a tabletop.

Scientists think that the dwarf planet crashed into a larger object long ago, which may have caused Haumea to rotate more quickly. The crash may have also created Haumea's two moons, Hi'iaka and Namaka. This dwarf planet takes a long time to complete one trip around the Sun—284 years.

Haumea is basically a giant rock with a thick coating of ice. Its frozen surface shines brightly, making it the third brightest object in the Kuiper Belt. With a good telescope, it is possible to see Haumea on a clear night. Scientists have seen a dark red spot on the dwarf planet's surface. They think this spot could contain important minerals. There is no coat thick enough to keep you warm on Haumea. The average temperature is a chilly -402 degrees Fahrenheit!

Two teams of astronomers discovered Haumea within a few months of each other in 2004–2005. It was the fifth dwarf planet to be identified. In 2017, scientists discovered that Haumea has rings! Astronomers were able to see them when the dwarf planet moved in front of a star. This means Haumea is the farthest object from the Sun with rings in our solar system.

A DEEPER LOOK

The Kuiper Belt is named after scientist Gerard Kuiper, pronounced *ky-purr*. Kuiper wanted to know where comets with small orbits came from. In 1951, he had an idea that there could be an area of icy objects beyond Neptune. He thought that was where the comets came from. No one had ever seen an area like this and Kuiper couldn't prove his theory, but he turned out to be right. That's why the belt is named after him! Scientists think that the icy objects are left over from when the solar system was formed. They could have formed a ninth planet, but Neptune's gravity kept them moving.

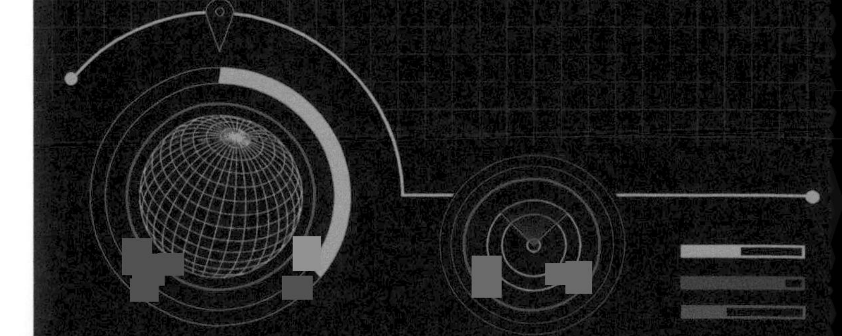

MAKEMAKE

DIAMETER:
888 MILES

DISTANCE FROM SUN:
4.3 BILLION MILES

DISTANCE FROM EARTH:
4.2 BILLION MILES
(AVERAGE CLOSEST DISTANCE)

KNOWN MOONS:
1 (MIGHT BE TEMPORARY)

LENGTH OF 1 DAY:
22.48 HOURS

LENGTH OF 1 YEAR:
310 EARTH YEARS

AVERAGE TEMPERATURE:
-406 DEGREES FAHRENHEIT

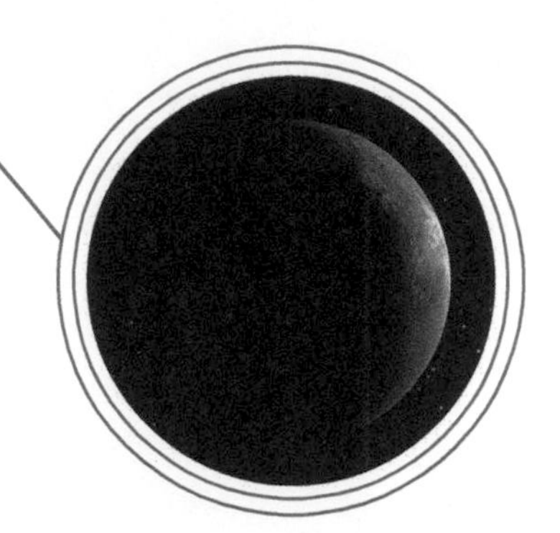

Also making its way around the Sun in the Kuiper Belt is Makemake, the second brightest object in the belt. Only Pluto shines brighter. This dwarf planet is a little smaller than Pluto and has an important place in space history. The discovery of Makemake and dwarf planet Eris is what helped astronomers decide on the difference between planets and dwarf planets.

Makemake is so far away from the Sun that it takes sunlight more than six hours to reach it. Each dimly lit day on the dwarf planet is just a little shorter than Earth's: about 22.48 hours long. Makemake's year is a lot longer though. The dwarf planet goes around the Sun only once every 310 years!

As with all objects in the Kuiper Belt, Makemake's distance from Earth makes it difficult for scientists to learn details about it. The dwarf planet seems to be a reddish-brown color, much like Pluto. It also seems to have frozen methane and ethane gases on its surface. Scientists think that frozen methane pellets the size of large peas may also be present.

Scientists think Makemake may have a very thin atmosphere made mostly up of a gas called nitrogen—but only when the dwarf planet is closer to the Sun. Its atmosphere escapes into space when Makemake leaves the Sun's warmth. Astronomers discovered one tiny moon orbiting the dwarf planet in 2016. It was found by NASA's Hubble Space Telescope. The moon is a dark charcoal color and only about 105 miles across. A car could drive across it in less than two hours!

Makemake was discovered in March 2005 at the Palomar Observatory in California. The team jokingly called it "Easterbunny" because they found it only a few days after Easter. A while later, the dwarf planet was renamed Makemake after a mythological god who created humankind. It was classified as a dwarf planet in 2008.

ERIS

DIAMETER:
1,445 MILES

DISTANCE FROM SUN:
6.3 BILLION MILES

DISTANCE FROM EARTH:
6.3 BILLION MILES
(AVERAGE CLOSEST DISTANCE)

KNOWN MOONS: 1

LENGTH OF 1 DAY:
25.9 HOURS

LENGTH OF 1 YEAR:
557 EARTH YEARS

AVERAGE TEMPERATURE:
-359 TO -469 DEGREES FAHRENHEIT

Let's check out the last dwarf planet in our solar system. We need to travel more than two billion miles farther than Makemake, Pluto, and Haumea to find it! Its name is Eris. Like Pluto, this dwarf planet is about the same size as Earth's Moon. Eris changed space history. When scientists found Eris, they thought it was larger than Pluto. They were going to call it the tenth planet. This is one of the discoveries that started the conversation about what is and isn't a planet.

Eris was discovered in October 2013 by the same group of scientists who discovered Makemake. They nicknamed the dwarf planet "Xena" after a warrior princess on a popular TV series. Scientists liked the name because it began with an X as in "Planet X." Eris is now named after the Greek goddess of discord and strife, or arguments and fighting. Many people think the name is a good fit because Eris plays a big part in the disagreement about the definition of a planet.

Eris travels the most distance in our solar system. It takes Eris 557 Earth years to orbit the Sun once—longer than any other planet or dwarf planet. Eris's oval orbit takes it far beyond the edge of the Kuiper Belt. This dwarf planet is so far away from the Sun, that if the Sun's light went out, Eris wouldn't notice for about ten hours!

Needless to say, Eris is one of the coldest planets, or dwarf planets, in the solar system. When Eris gets too far from the Sun, its atmosphere freezes and falls to the dwarf planet's rocky white surface as snow. The atmosphere thaws as it moves closer to the Sun again.

Eris has one very small moon called Dysnomia. It was named after the goddess Eris's daughter, Dysnomia, the demon goddess of lawlessness, or disorder. Dysnomia travels around Eris once every 16 days. Scientists were glad to discover this moon because they can measure the mass of Eris by observing how long Dysnomia takes to orbit.

EXPLORE MORE!

Visit NASA's website, solarsystem.nasa.gov/planets, to look at 3-D models of the planets and dwarf planets in our solar system. Click and drag the mouse on the models to see all the different parts of each planet. It's almost as good as seeing them in real life!

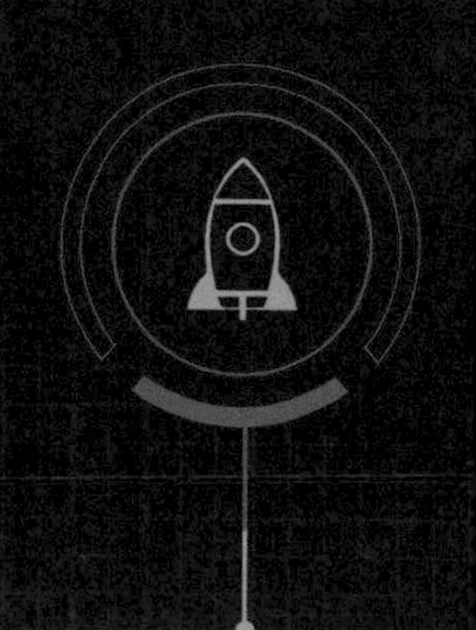

Chapter Six

EXPLORING SPACE

Has this book made you curious to learn more about space? Great! There are many ways to learn new things. You can observe the night sky with your eyes, borrow books from your school or local library, complete the activities in this book, visit a planetarium, or use binoculars or a telescope to see objects in space more up close. This illustration shows an optical telescope, which scientists use to view things very far away.

I encourage you to take time to learn more about our solar system and universe. Learning about space is a never-ending process that can lead to exciting discoveries! Who knows? Someday *you* might find a brand-new planet or star!

Stargazing

One way you can explore more from home is by looking for **constellations** in the night sky. Humans are very good at finding patterns, and constellations are easier patterns to recognize than single stars. The outlines of people, animals, and objects helped astronomers all over the world make maps of the night skies. Constellations even help people sail ships and create calendars.

Today, astronomers use 88 official constellations to divide the night sky into regions. Most of the names come from what Greeks and Romans called them thousands of years ago. Cultures around the world have different numbers of constellations and different names for them. We still use many of the names today. Some of the names are more than 4,000 years old!

Let's check out some constellations! Go outside on a clear, dark night with a flashlight. Cover your flashlight with red cellophane since bright white lights can ruin your night vision. Look up at the sky. You can search for constellations by thinking of them as dot-to-dot drawings. When you first look up, the sky can seem very busy, but concentrate on finding the brightest stars. You can use a round constellation map, or planisphere, to identify the stars that are part of constellations. The constellations you see will depend on where you live, what time it is, and the season.

URSA MAJOR
(THE BIG DIPPER)

Ursa Major, the Great Bear constellation, is where you'll find the Big Dipper. This constellation is most visible in the spring when it's high above the northeast horizon. The Big Dipper is one of the easiest groups of stars to see. It is made up of seven bright stars at the center of the bear, which form a "dipper" (ladle or spoon). It can be difficult to see all of the bear, especially if you are surrounded by light from buildings.

URSA MINOR
(THE LITTLE DIPPER)

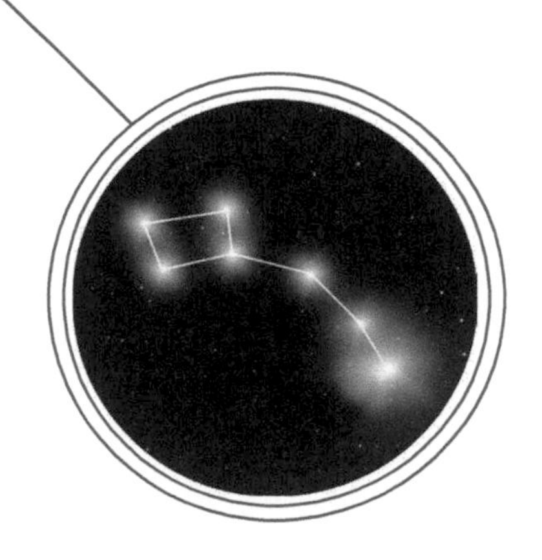

Ursa Minor, the Little Bear constellation, is where you'll find the Little Dipper. If the Big Dipper is right-side up in the sky, then the Little Dipper is upside down, and vice versa. The handles of each constellation point in opposite directions. Polaris, the North Star, is at the end of the Little Dipper's handle. The Little Dipper's stars are faint compared with the Big Dipper's stars so you need a dark, clear night to have the best chance of finding it.

ORION

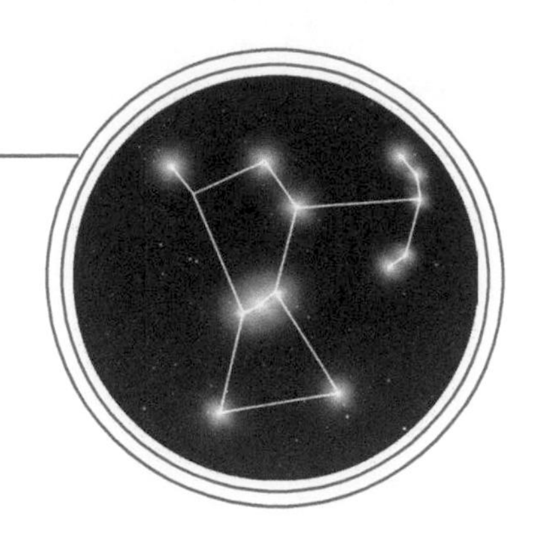

According to Greek mythology, Orion was the son of Neptune and a great hunter. After Orion bragged that he would hunt down every animal in the world, Gaia, the earth goddess, sent a scorpion to sting and kill him. To find the constellation Orion, look for the three stars in a line that form his belt. They point down to Sirius, the brightest star in the night sky. The reddish star in Orion's shoulder is named Betelgeuse. It is a red supergiant that will explode in a supernova some day!

TAURUS

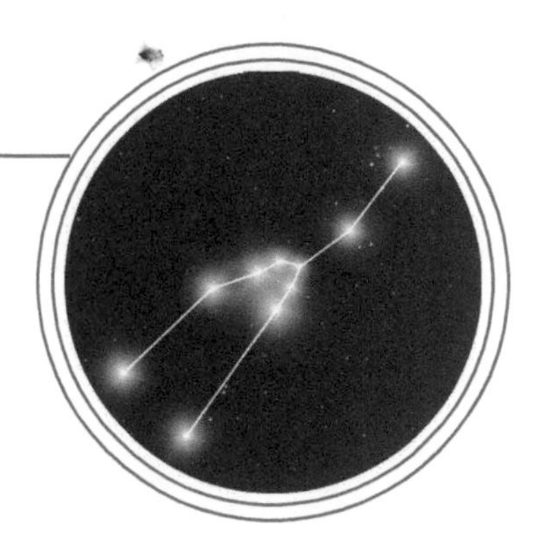

The constellation Taurus is also called "the Bull." Taurus was discovered long ago in the early Bronze Age. It is one of the most visible constellations. The bull charges

EXPLORE MORE!

Different people see different things when they look at the stars. After looking for the named constellations, see what animals, people, or objects you can find. Sketch the stars you see and the shapes you imagine, and give your constellation a name!

through the sky from November until March. Your best chance of seeing it is in January. Aldebaran is a famous red giant star and is part of the Bull's face. Taurus also contains a cluster of stars called the Pleiades.

GEMINI

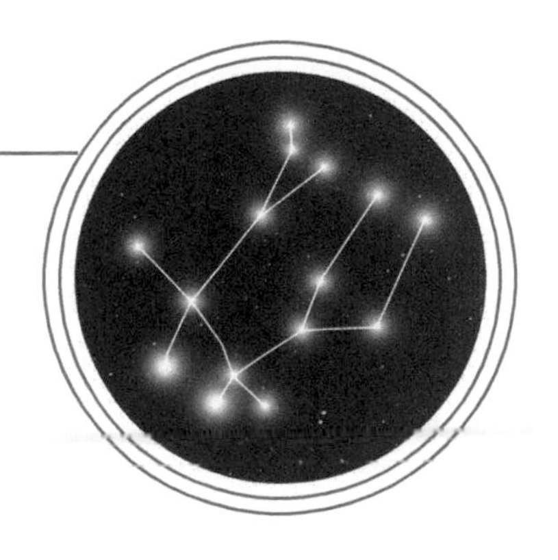

Gemini is the Latin word for "twins." Gemini is pretty easy to spot in the sky. Once you've found Taurus, look to the east for two bright stars. These are the heads of the twins. February is your best chance at a glimpse. If you are looking in April or May, you can see Gemini in the west soon after sunset.

Looking for Planets

When you're looking at the night sky, both planets and stars look like stars. How can you tell the difference? The easiest way is to remember the lullaby "Twinkle, Twinkle, Little Star." The song doesn't say "Little Planet" because only stars seem to twinkle.

Stars are many light-years away and look like tiny points of light. When starlight travels to Earth, it moves through the atmosphere and then to your eye. The atmosphere has lots of moving air and gas, which blurs the star's light and makes it seem to twinkle. Planets are much closer to Earth and, through binoculars or a small telescope, look like small circles, or disks. Because of this, planets don't twinkle when you look at them. They are also usually as bright or brighter than stars.

You can see five planets from Earth without a telescope: Mercury, Venus, Mars, Jupiter, and Saturn. You can see them at different times of year. It is often best to look for planets when it is just beginning to get dark (dusk) or when it is starting to get light again (dawn). Mercury and Venus will always closely follow the Sun in the sky, while Mars, Jupiter, and Saturn may be farther away. Before you go outside to look, visit a website like www.earthsky.org to learn which planets you can see that night and where to look for them.

MERCURY: Look to the horizon at sunset or sunrise for Mercury. It may have a yellow color to it.

VENUS: Silvery Venus is so bright it's sometimes mistaken for an airplane! It is one of the easiest planets to spot.

MARS: When Mars is close to Earth, the planet can look rather bright and have a reddish color.

JUPITER: Jupiter is usually the second-brightest planet in the sky behind Venus. It is also one of the easiest planets to spot.

SATURN: Saturn shines with a golden color and can look a lot like a star at first. If possible, you should look at it with a telescope to see its spectacular wide rings.

What Else Can We See?

Besides stars and planets, there are many other things to view in the night (and sometimes day) sky. Lunar eclipses can easily be viewed at night when the timing is right. Solar eclipses are rare, but if you're lucky enough to see one, make sure you follow the guidelines for viewing it safely. You can also look for meteor showers and comets, like the one in this illustration. You may even be able to see the

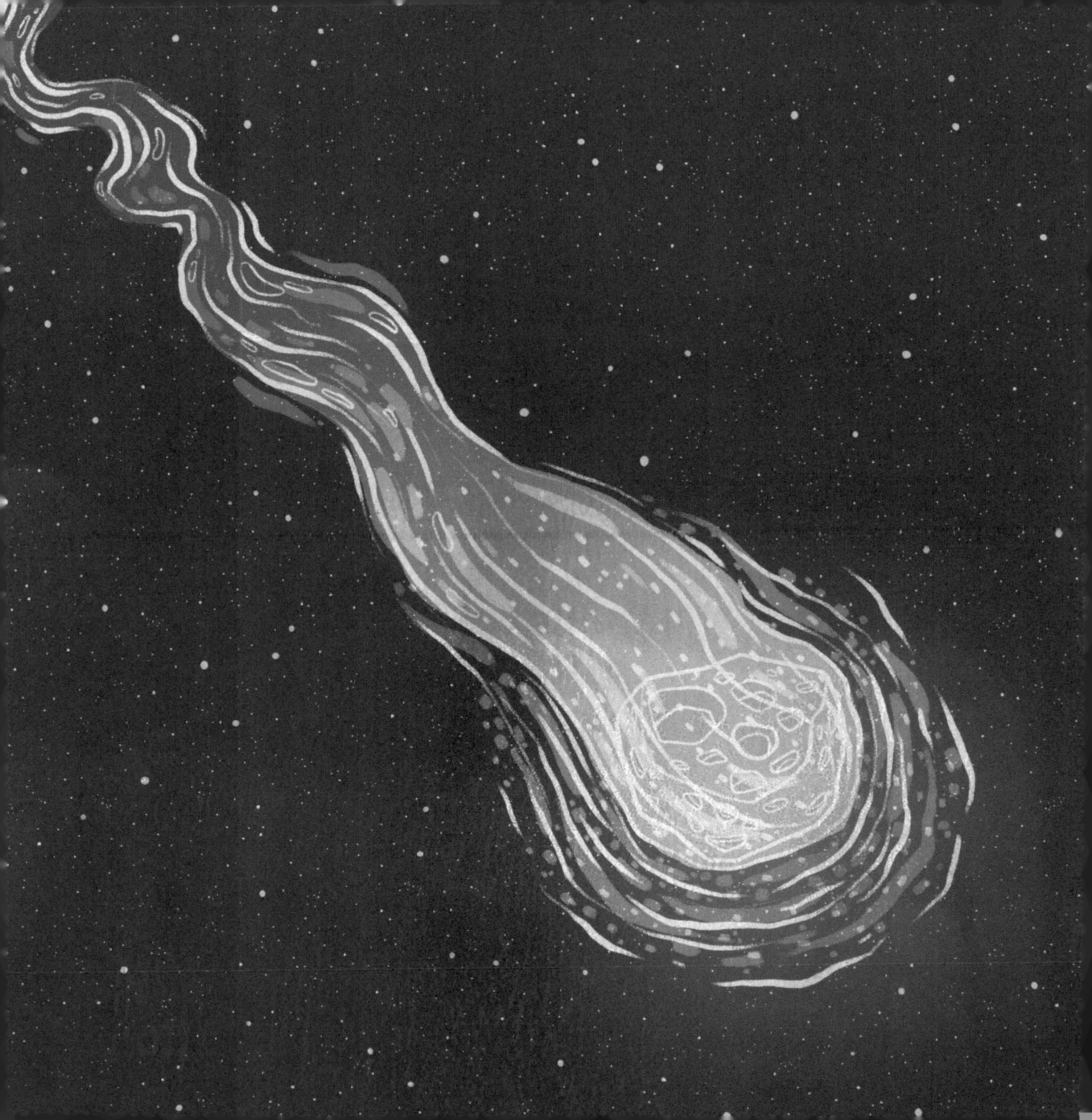

International Space Station zipping across the sky or satellites slowly orbiting in space. If you see a bright, steadily moving light that doesn't flash like an airplane, chances are you've found a satellite. You can visit heavens-above.com to learn which satellites are visible from your location.

COMETS

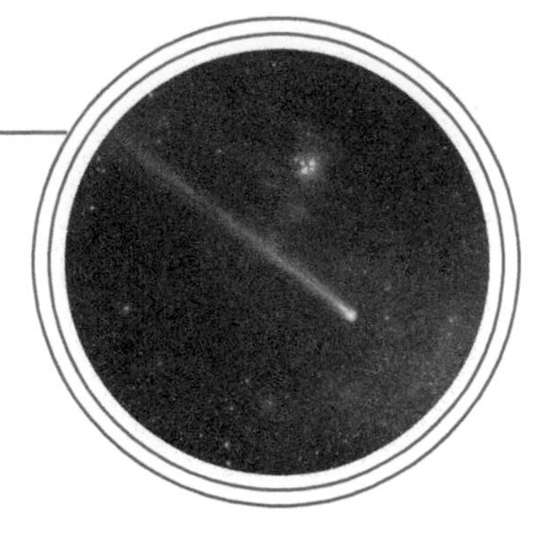

What are **comets**? They are balls of ice that orbit the Sun. Comets are made up of mostly frozen methane gas, ammonia, and water, so they have been nicknamed "dirty snowballs." Most comets follow an orbit far from Earth, but a few come close enough to be seen without a telescope.

When a comet gets close to the Sun, it's easier to see. As the Sun warms the comet, the comet's ice starts to melt and mixes with its gasses, forming a cloud of gas and dust around the snowball. The cloud gets spread out by the effects of the Sun into (usually two) long glowing tails. Comet tails can be several million miles long!

Astronomers think there could be billions of comets orbiting the Sun. They know of about 6,500 comets right now. On average, a comet is visible with just your eyes once every 10 years.

Halley's Comet is one of the most well-known comets. It comes near the Sun and Earth about every 75 years. Next time a comet visits, head outside on a clear night with binoculars or a small telescope to try and spot one of these truly incredible sights!

Another thing you can see in the night sky is a meteor shower, like the one in this illustration. Turn the page to learn more!

A DEEPER LOOK

Did you know that some meteoroids make it to the ground without burning up completely? A meteorite is a meteoroid that has landed on Earth.

Most meteorites are the size of tiny dust particles. Millions of meteorite specks fall to Earth every single day—many hundreds of tons' worth! When you come inside after being outside, you might be bringing a little bit of space dust with you!

A few meteorites are as big as boulders. The biggest meteorite ever discovered on Earth's surface is called Hoba because it was found on a farm in Namibia named Hoba West. Hoba weighs 66 tons—as much as 11 elephants—and is about 9 feet long!

METEOR SHOWERS

Meteors are streaks of light in the night sky caused by meteoroids hitting the Earth's atmosphere. Meteoroids can be as small as a piece of dust, as big as a giant boulder, or any size in between. As a meteoroid falls toward Earth, the rock becomes very hot and burns up. This is because of friction, or the rock "rubbing" against the air as it falls. When we see this happen, we often say we saw a "shooting star" even though meteors have nothing to do with stars!

If a group of meteoroids makes its way to Earth or when Earth orbits through a large group of meteoroids, it's called a "meteor shower." Meteor showers sometimes happen when Earth passes through a comet's tail. During these showers, you can expect to see many meteors in a short amount of time.

Plan ahead if you would like to view a meteor shower. One of the most well-known showers happens in August and is called the Perseids. The best time to watch a meteor shower is before the Sun comes up, when the Moon is not yet out. While you can see the most meteors during a meteor shower, you don't have to wait for a meteor shower to see just one or two. They are a common sight on most nights.

A DEEPER LOOK

Have you ever wondered how astronomers can predict a meteor shower? The answer is really quite simple.

Comets and asteroids leave a trail of icy fragments or pieces of space dust as they zip past Earth. These pieces stay in a band that stretches across Earth's orbit. When Earth passes through the band of particles, a meteor shower occurs, which usually happens only once per orbit. Because Earth passes through the same band at the same time every year, astronomers can easily predict when the next meteor shower will be.

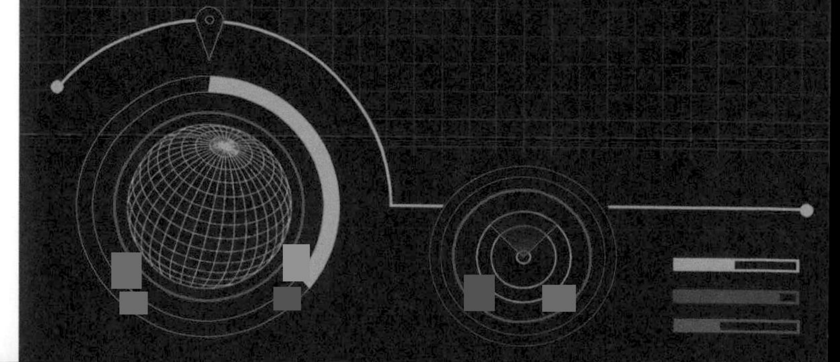

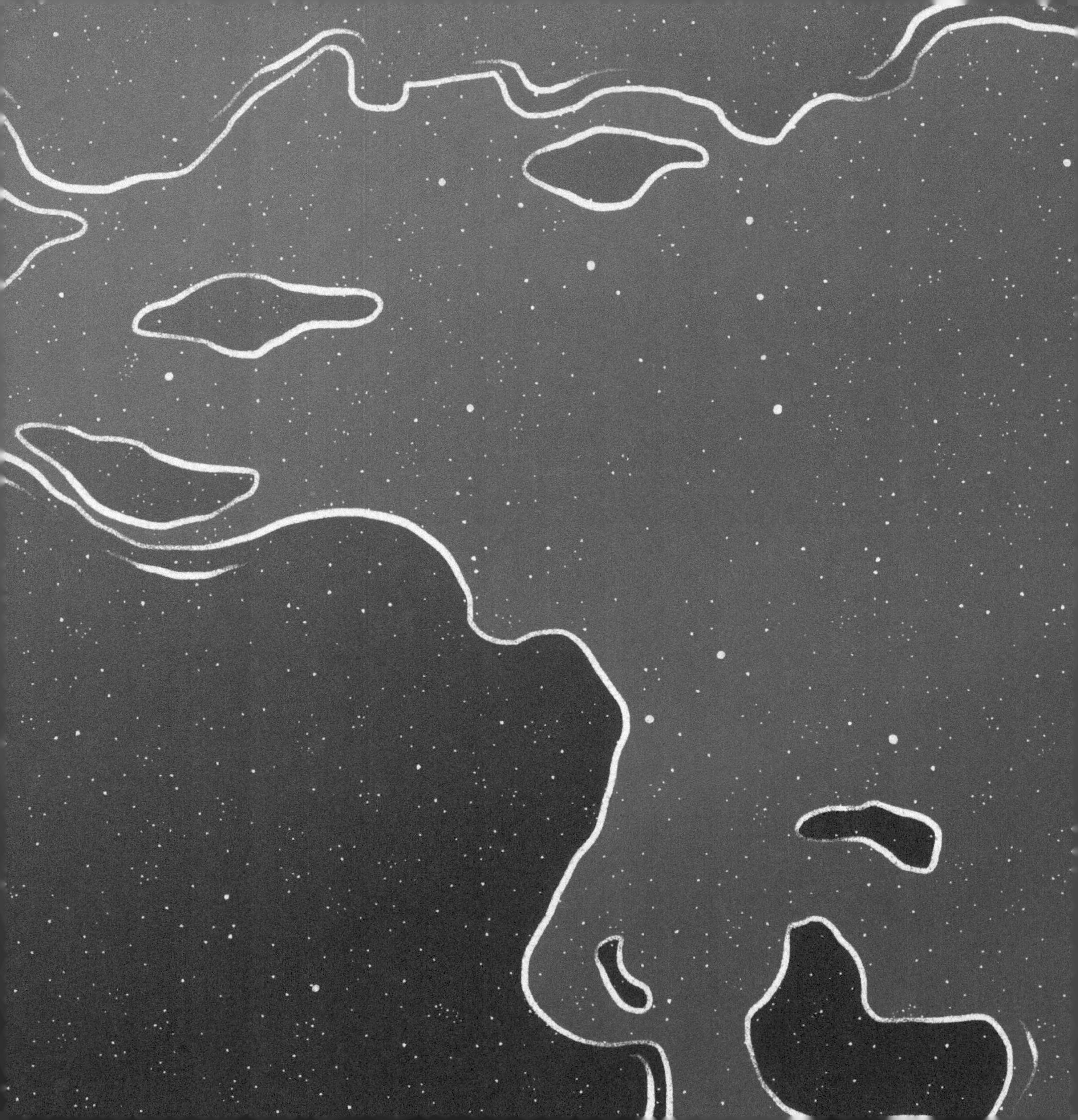

The Mystery Continues

We know a lot about our solar system and outer space, but there is still so much for scientists to figure out. Much of outer space is a mystery, but the more we find out about space, the more curious we become. Hopefully, this book has made you want to learn more.

I encourage you to be a lifelong learner. Now that you know how to look for constellations and planets, you are ready to study the night sky with binoculars, a telescope, or just your eyes. Do you want to be an astronaut or a scientist? Do you dream of looking at Earth from the Moon?

It's never too early to begin your journey. Visit planetariums and space museums. Read every book about space that interests you. You could even join an astronomy club! Ask family members and teachers to help you learn about space using a computer or tablet. Visit NASA's website for kids (spaceplace.nasa.gov/en/search/kids/).

Keep up with news about space. Follow along as space probes and spaceships explore distant outer space.

You never know where your journey may lead you! Someday you could be the person who visits Mars, or discovers a brand-new planet. So, reach for the stars. They are waiting for you!

Glossary

asteroid: A small object made of rock or metal that travels through space.

astronomy: The scientific study of celestial objects (for example, stars and planets) in the universe.

atmosphere: The layer of gases held by gravity to a planet, moon, or other object in space.

axis: An imaginary line that passes through the center of a star, moon, or planet around which the object rotates or spins.

black hole: This is an object that has so much gravity that it pulls everything into itself. Black holes are so strong that not even light can escape. That is why we call them black.

comet: A ball of frozen water, gas, and rock that orbits the Sun and is often referred to as a "dirty snowball."

constellation: A group of stars that forms a pattern in the night sky. There are currently 88 of them.

dark energy: This mysterious force is making the universe expand. Most of the universe is made of dark energy.

dark matter: The stuff in space that has gravity but can't be seen with a telescope and is still a mystery to modern astronomy.

dwarf planet: A large, round object that orbits the Sun but hasn't cleared away other objects in its orbit. It is typically much smaller than a planet.

galaxy: A collection of billions (or even trillions) of stars held together in one group by gravity. Our galaxy is the Milky Way. Stars in galaxies may have their own solar systems.

gravity: The force that pulls one object to another.

light-year: The distance traveled by a particle of light, moving at the speed of light, in one Earth year—about 5.8 trillion (5,800,000,000,000) miles.

lunar eclipse: An event that occurs when the Moon moves exactly on the opposite side of the Earth from the Sun. The Moon moves into Earth's shadow during this time. The opposite of a **solar eclipse**.

mantle: The part of a planet that is between its surface (crust) and core (middle).

matter: Anything that takes up space and has weight.

meteor: A streak of light in the sky caused by a meteoroid (space dust and rock) falling through Earth's atmosphere at high speed. A meteor is sometimes called a shooting star.

moon: A natural object that orbits a planetary body.

nebula: This is a large space cloud made of dust or gas. Stars are born in nebulae, the plural form of nebula.

oblate spheroid: A planet or other round space object that is not perfectly round, but is "squashed," so it bulges in the middle.

observable universe: The part of the universe that can be seen from Earth or from telescopes in space.

orbit: The path an object follows as it travels around another object. Earth's orbit around the Sun is relatively circular.

planet: A large, round object that orbits the Sun and has cleared away all other objects in its orbit.

planetary system: A group of objects that orbits around a star or star system. Our planetary system is called the solar system.

probe: A spacecraft that is sent out to learn things about an object in space. There are no people on probes.

prominence: A giant flaming loop or arm that the Sun throws into space from its surface.

satellite: A spacecraft or other human-made object that orbits a larger object.

solar eclipse: This event occurs when the Moon is exactly between the Earth and the Sun, blocking the Sun's rays and shadowing the Earth. The opposite of a **lunar eclipse**.

solar system: Our planetary system, including the Sun, and all the planets, moons, and other objects that orbit it.

star: A giant ball of gas that makes its own light and heat.

terrestrial planets: The four dense, rocky planets closest to the Sun.

universe: All energy and matter, including Earth, objects in space, vast galaxies, and stars, and everything in between.

More to Explore

Planet Earth by Christine Taylor-Butler

The Secret Galaxy by Fran Hodgkins

Space!: The Universe as You've Never Seen It Before by DK

Super Cool Space Facts: A Fun, Fact-filled Space Book for Kids by Bruce Betts

13 Planets: The Latest View of the Solar System by David A. Aguilar

NASA, www.nasa.gov

NASA Kids' Club, www.nasa.gov/kidsclub

Index

U

V

W

Y

About the Author

is a kindergarten teacher in Middle Tennessee with well over a decade of experience at the school she attended as a young child. She is the author of Pencils to Pigtails, a popular blog for parents and teachers. Hilary is also a featured presenter at regional and national educational events held around the country. She holds a master's degree in Instructional Leadership and enjoys coordinating and teaching her school district's free English class for parents. Her favorite days are spent with her two young girls, husband, parents, siblings, and the rest of her very large extended family. You can connect with Hilary on all social media platforms using the handle @pencilstopigtails.

Printed in the USA
CPSIA information can be obtained
at www.ICGtesting.com
LVHW051930091223
765801LV00003B/31

9 781646 119288